Fundamentals of Conceptual Physics

Fundamentals of Conceptual Physics

Leonel Ware

STATES
ACADEMIC PRESS
www.statesacademicpress.com

Published by States Academic Press
109 South 5th Street,
Brooklyn, NY 11249, USA
www.statesacademicpress.com

Fundamentals of Conceptual Physics
Leonel Ware

International Standard Book Number: 978-1-63989-225-9 (Hardback)

Cataloging-in-Publication Data

Fundamentals of conceptual physics / Leonel Ware.
p. cm.
Includes bibliographical references and index.
ISBN 978-1-63989-225-9
1. Physics. 2. Dynamics. 3. Physical sciences. I. Ware, Leonel.
QC21.3 .C66 2022
530--dc23

Contents

Preface

Physics is a natural science which deals with the study of matter, its motion and behavior through space and time along with the related entities such as energy and force. The primary objective of this discipline is to understand the behavior of the universe. Conceptual physics is one of the approaches towards teaching physics which places emphasis on the concept and ideas of physics rather than its mathematical aspects. Physics is broadly classified into classical physics and modern physics. Subjects such as thermodynamics, electromagnetics, calculus, planetary motion are studied under classical physics. Modern physics is concerned with the phenomena which are governed by the laws of quantum mechanics and theory of relativity. Some major areas of research under this field are nuclear and particle physics, atomic physics, astrophysics, applied physics and condensed matter physics. The various sub-fields of physics along with technological progress that have future implications are glanced at in this book. Different approaches, evaluations and methodologies and advanced studies on physics have been included herein. This book is appropriate for those seeking detailed information in this area.

A foreword of all Chapters of the book is provided below:

Chapter 1 - The branch of science which is concerned with the study of motion and behavior of matter with respect to space and time is known as physics. There are several significant aspects of physics such as mechanics, momentum, motion, force, energy, heat, etc. This is an introductory chapter which will briefly introduce all these facets of physics; **Chapter 2** - Classical mechanics is a theory of physics that explains the motion of objects under the influence of forces. Some of the important topics studied under classical mechanics are Hamilton's principle and Lagrange's equation, rotating frame and motion of a rigid body. This chapter closely examines these key concepts of classical mechanics to provide an extensive understanding of the subject; **Chapter 3** - Quantum mechanics is a fundamental theory of physics that describes physical properties of nature at atomic and subatomic scale. The various principles and equations studied in relation to quantum mechanics are Heisenberg's uncertainty principle, generalized uncertainty principle, Heisenberg equations of motion, etc. These topics are crucial for a complete understanding of the subject; **Chapter 4** - The branch of mechanics in which statistical methods and probability theory are applied to describe the behavior of a large number of atoms is known as statistical mechanics. A few of the topics studied under this domain are statistical thermodynamics, statistical ensemble of thermodynamics, classical mechanical formulation of statistical mechanics, etc. All these diverse principles of statistical mechanics have been carefully analyzed in this chapter; **Chapter 5** - The branch of mechanics that is compatible with special relativity and general relativity is known as relativistic mechanics. It deals with the motion of bodies whose relative velocities are equal to the speed of light. This chapter discusses in detail the theories and methodologies related to relativistic mechanics such as theory of relativity, relativistic space-time, relativistic momentum, etc; **Chapter 6** - Physics can be understood through various theories, principles and laws such as classical Hamiltonian mechanics, Archimedes' principle, Bernoulli's principle, physical law, conservation law, inverse-square law, Stokes' law, Faxen's law and Hagen-Poiseuille equation. This chapter has been carefully written to provide an easy understanding of these various theories, principles and laws of physics.

I would like to thank the entire editorial team who made sincere efforts for this book and my family who supported me in my efforts of working on this book. I take this opportunity to thank all those who have been a guiding force throughout my life.

Leonel Ware

1

Fundamentals of Physics

The branch of science which is concerned with the study of motion and behavior of matter with respect to space and time is known as physics. There are several significant aspects of physics such as mechanics, momentum, motion, force, energy, heat, etc. This is an introductory chapter which will briefly introduce all these facets of physics.

Mechanics

Mechanics is the branch of Physics dealing with the study of motion when subjected to forces or displacements, and the subsequent effects of the bodies on their environment.

There are two sub disciplines: Classical Fields and Quantum Fields.

Classical Mechanics

Classical mechanics deals with the question of how an object moves when it is subjected to various forces, and also with the question of what forces act on an object which is not moving. The word "classical" indicates that we are not discussing situations in which an object moves with a velocity which is an appreciable fraction of the velocity of light or phenomena on the atomic scale. The description of atomic phenomena requires quantum mechanics, and the description of phenomena at very high velocities requires Einstein's Theory of Relativity.

Both quantum mechanics and relativity were invented in the twentieth century; the laws of classical mechanics were stated by Sir Isaac Newton in 1687. The laws of classical mechanics enable us to calculate the trajectories of bullets and baseballs, planets and space vehicles. Using these laws we can predict the position-versus-time relation for a cylinder rolling down an inclined boat or for an oscillating pendulum and can calculate the tension in the wire when a picture is hanging on a wall.

The practical importance of the subject hardly requires demonstration in a world which contains buildings, automobiles, airplanes, ballistic missiles, bridges, for a person who does not have any professional reason there is a compelling intellectual reason to study classical mechanics: this is the example par excellence of a theory which explains an incredible multitude of phenomena on the basis of a minimal number of simple principles.

Quantum Mechanics

Quantum mechanics can be thought of roughly as the study of physics on very small length scales, although there are also certain macroscopic systems it directly applies to. The descriptor "quantum" arises because, in contrast with classical mechanics, certain quantities take on only discrete values. However, some quantities still take on continuous values, in quantum mechanics, particles have wavelike properties, and a particular wave equation, the Schrodinger equation, governs how these waves behave.

In some respect, quantum mechanics is just another example of a system governed by a wave equation. However, although it is fairly straightforward to deal with the actual waves, there are many things about quantum mechanics that are a combination of subtle, perplexing, and bizarre. To name a few: the measurement problem, hidden variables along with Bell's theorem, and wave-particle duality. Even though there are many things that are highly confusing about quantum mechanics, the nice thing is that it's relatively easy to apply quantum mechanics to a physical system to figure out how it behaves.

Statistical Mechanics

This branch of statistical mechanics which treats and extends classical thermodynamics is known as statistical thermodynamics or equilibrium statistical mechanics.

We apply statistical mechanics to solve for real systems (a system for many particles). We can easily solve the Schrodinger's equation for 1 particle, atom or a molecule. For many particles, the solution will take the form:

total = linear combination of a(1) b(2) c(3)

Where a means a particle in state a with an energy E_a. For example, we consider particles conned in a cubic box of length L. From quantum mechanics, the possible energies for each particle is:

E= 22m2L2(n2x+n2y+n2z)

For example in an ideal gas, we assume that the molecules are non-interacting, i.e. they do not affect each other's energy levels. Each particle contains certain energy. At $T > 0$, the system possesses total energy, E.

So, how is E distributed among the particles? Another question would be, how the particles are distributed over the energy levels. We apply statistical mechanics to provide the answer and thermodynamics demands that the entropy be maximum at equilibrium.

Thermodynamics is concerned about heat and the direction of heat flow, whereas statistical mechanics gives a microscopic perspective of heat in terms of the structure of matter and provides a way of evaluating the thermal properties of matter, for e.g., heat Capacity.

Momentum

Momentum is a product of the mass of a particle and its velocity. Momentum is a vector quantity; i.e., it has both magnitude and direction. Isaac Newton's second law of motion states that the time rate of change of momentum is equal to the force acting on the particle.

From Newton's second law it follows that, if a constant force acts on a particle for a given time, the product of force and the time interval (the impulse) is equal to the change in the momentum. Conversely, the momentum of a particle is a measure of the time required for a constant force to bring it to rest.

The momentum of any collection of particles is equal to the vector sum of the individual momenta. According to Newton's third law, the particles exert equal and opposite forces on one another, so any change in the momentum of one particle is exactly balanced by an equal and opposite change of the momentum of another particle. Thus, in the absence of a net external force acting on a collection of particles, their total momentum never changes; this is the meaning of the law of conservation of momentum.

Motion

In physics, the motion is the change in position of an object with respect to its surroundings in a given interval of time. The motion of an object with some mass can be described in terms of the following:

- Distance
- Displacement
- Speed
- Velocity
- Time
- Acceleration

Types of Motion in Physics

Motion of an object depends on the type of force acting on the body. Examples of different kinds of motion are given below:

- Translational: It is the type, where an object moves along a path in any of the three dimensions.

- Rotational: It is the type, where an object moves along a circular path about a fixed axis.
- Linear: It is a type of translational motion where the body moves in a single direction along a single dimension.
- Periodic: It is the type of motion that repeats itself after certain intervals of time.
- Simple Harmonic: It is the type of motion like that of a simple pendulum where a restoring force acts in the direction opposite to the direction of motion of the object. This restoring force is proportional to the displacement of the object from mean position.
- Projectile: It is the type of motion which has a horizontal displacement as well as vertical displacement.
- Oscillatory: It is the type of motion which is repetitive in nature within a time frame. If it is mechanical in nature it is called vibration.

Laws of Motion

Newton's Laws of Motion laid the foundation for classical mechanics today. Although subject to minor limitations, these laws of motion are valid everywhere and are therefore used. The laws are given as stated below in a brief description:

- First Law: Any object will remain in its existing state of motion or rest unless a net external force acts on it.
- Second Law: If an object has a certain mass, greater the mass of this object, greater will the force required be to accelerate the object. It is represented by the equation $F = ma$, where 'F' is the force on the object, 'm' is the mass of the object and 'a' is the acceleration of the object.
- Third Law: For every action, there is an equal and opposite reaction.

Force

Force is any action that tends to maintain or alter the motion of a body or to distort it. The concept of force is commonly explained in terms of Isaac Newton's three laws of motion set forth in his Principia Mathematica. According to Newton's first principle, a body that is at rest or moving at a uniform rate in a straight line will remain in that state until some force is applied to it. The second law says that when an external force acts on a body, it produces acceleration (change in velocity) of the body in the direction of the force. The magnitude of the acceleration is directly proportional to the magnitude of the external force and inversely proportional to the quantity of matter in the body. Newton's third law states that when one body exerts a force on another body, the second body exerts an equal force on the first body. This principle of action and reaction explains why a force tends to deform a body (i.e., change its shape) whether or not it causes the body to move. The deformation of a body can usually be neglected when investigating its motion.

Because force has both magnitude and direction, it is a vector quantity. The representation of forces by vectors implies that they are concentrated either at a single point or along a single line. This is, however, physically impossible. On a loaded component of a structure, for example, the applied force produces an internal force, or stress, that is distributed over the cross section of the component. The force of gravity is invariably distributed throughout the volume of a body. Nonetheless, when the equilibrium of a body is the primary consideration, it is generally valid as well as convenient to assume that the forces are concentrated at a single point. In the case of gravitational force, the total weight of a body may be assumed to be concentrated at its centre of gravity.

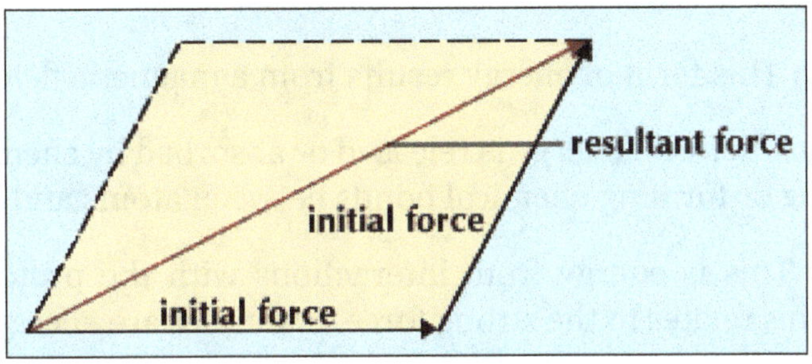

Two forces applied simultaneously to the same point have the same effect as a single equivalent force. The resultant force can be found by constructing a parallelogram with the initial force vectors forming two adjacent sides. The diagonal of the parallelogram gives the resultant force vector.

Physicists use the newton, a unit of the International System (SI), for measuring force. A newton is the force needed to accelerate a body weighing one kilogram by one metre per second per second. The formula F = ma is employed to calculate the number of newton's required to increase or decrease the velocity of a given body. In countries still using the English system of measurement, engineers commonly measure force in pounds. One pound of force imparts to a one-pound object an acceleration of 32.17 feet per second squared.

Energy

Energy is defined as the capacity of a physical system to perform work. However, it's important to keep in mind that just because energy exists, that doesn't mean it's necessarily available to do work.

Forms of Energy

Energy exists in several forms such as heat, kinetic or mechanical energy, light, potential energy, and electrical energy:

- Heat: Heat or thermal energy is energy from the movement of atoms or molecules. It may be considered as energy relating to temperature.
- Kinetic Energy: Kinetic energy is the energy of motion. A swinging pendulum has kinetic energy.

- Potential Energy: This is energy due to an object's position. For example, a ball sitting on a table has potential energy with respect to the floor because gravity acts upon it.
- Mechanical Energy: Mechanical energy is the sum of the kinetic and potential energy of a body.
- Light: Photons are a form of energy.
- Electrical Energy: This is energy from the movement of charged particles, such as protons, electrons, or ions.
- Magnetic Energy: This form of energy results from a magnetic field.
- Chemical Energy: Chemical energy is released or absorbed by chemical reactions. It is produced by breaking or forming chemical bonds between atoms and molecules.
- Nuclear Energy: This is energy from interactions with the protons and neutrons of an atom. Typically this relates to the strong force. Examples are energy released by fission and fusion.

Other forms of energy may include geothermal energy and classification of energy as renewable or nonrenewable. There may be overlap between forms of energy and an object invariably possesses more than one type at a time. For example, a swinging pendulum has both kinetic and potential energy, thermal energy, and (depending on its composition) may have electrical and magnetic energy.

Law of Conservation of Energy

According to the law of conservation of energy, the total energy of a system remains constant, though energy may transform into another form. Two billiard balls colliding, for example, may come to rest, with the resulting energy becoming sound and perhaps a bit of heat at the point of collision. When the balls are in motion, they have kinetic energy. Whether they are in motion or stationary, they also have potential energy because they are on a table above the ground.

Energy cannot be created, nor destroyed, but it can change forms and is also related to mass. The mass-energy equivalence theory states an object at rest in a frame of reference has a rest energy. If additional energy is supplied to the object, it actually increases that object's mass. For example, if you heat a steel bearing (adding thermal energy), you very slightly increase its mass.

Heat

Heat is a form of energy that is transferred from one body to another as the result of a difference in temperature. If two bodies at different temperatures are brought together, energy is transferred—i.e., heat flows—from the hotter body to the colder. The effect of this transfers of energy usually, but not always, is an increase in the temperature of the colder body and a decrease in the temperature of the hotter body. A substance may absorb heat without an increase in temperature by changing from one physical state (or phase) to another, as from a solid to a liquid (melting), from a solid to a vapour (sublimation), from a liquid to a vapour (boiling), or from one solid form to another (usually called a crystalline transition). The important distinction between heat and temperature (heat being a form of energy and temperature a measure of the amount of that energy present in a body) was clarified during the 18th and 19th centuries.

Heat as a Form of Energy

Because all of the many forms of energy, including heat, can be converted into work, amounts of energy are expressed in units of work, such as joules, foot-pounds, kilowatt-hours, or calories. Exact relationships exist between the amounts of heat added to or removed from a body and the magnitude of the effects on the state of the body. The two units of heat most commonly used are the calorie and the British thermal unit (BTU). The calorie (or gram-calorie) is the amount of energy required to raise the temperature of one gram of water from 14.5 to 15.5 °C; the BTU is the amount of energy required to raise the temperature of one pound of water from 63 to 64 °F. One BTU is approximately 252 calories. Both definitions specify that the temperature changes are to be measured at a constant pressure of one atmosphere, because the amounts of energy involved depend in part on pressure. The calorie used in measuring the energy content of foods is the large calorie, or kilogram-calorie, equal to 1,000 gram-calories.

In general, the amount of energy required to raise a unit mass of a substance through a specified temperature interval is called the heat capacity, or the specific heat, of that substance. The quantity of energy necessary to raise the temperature of a body one degree varies depending upon the restraints imposed. If heat is added to a gas confined at constant volume, the amount of heat needed to cause a one-degree temperature rise is less than if the heat is added to the same gas free to expand (as in a cylinder fitted with a movable piston) and so do work. In the first case, all the energy goes into raising the temperature of the gas, but in the second case, the energy not only contributes to the temperature increase of the gas but also provides the energy necessary for the work done by the gas on the piston. Consequently, the specific heat of a substance depends on these conditions. The most commonly determined specific heats are the specific heat at constant volume and the specific heat at constant pressure. The heat capacities of many solid elements were shown to be closely related to their atomic weights by the French scientists Pierre-Louis Dulong and Alexis-Thérèse Petit in 1819. The so-called law of Dulong and Petit was useful in determining the atomic weights of certain metallic elements, but there are many exceptions to it; the deviations were later found to be explainable on the basis of quantum mechanics.

It is incorrect to speak of the heat in a body, because heat is restricted to energy being transferred. Energy stored in a body is not heat (nor is it work, as work is also energy in transit). It is customary, however, to speak of sensible and latent heat. The latent heat, also called the heat of vaporization, is the amount of energy necessary to change a liquid to a vapour at constant temperature and pressure. The energy required to melt a solid to a liquid is called the heat of fusion, and the heat of sublimation is the energy necessary to change a solid directly to a vapour, these changes also taking place under conditions of constant temperature and pressure.

Air is a mixture of gases and water vapour, and it is possible for the water present in the air to change phase; i.e., it may become liquid (rain) or solid (snow). To distinguish between the energy associated with the phase change (the latent heat) and the energy required for a temperature change, the concept of sensible heat was introduced. In a mixture of water vapour and air, the sensible heat is the energy necessary to produce a particular temperature change excluding any energy required for a phase change.

Gravitation

Gravitation or just gravity is the force of attraction between any two bodies. All the objects in the universe attract each other with a certain amount of force, but in most of the cases, the force is too weak to be observed due to the very large distance of separation. Besides, gravity's range is infinite but the effect becomes weaker as objects move away.

This force of attraction was first observed by Sir Isaac Newton and was presented as Newton's law of gravitation in the year 1680. However, gravitation can generally exist in two main instances:

- Gravitation may be the attraction of objects by the earth. Example: If a body (ball) is thrown upwards, it reaches a certain height and falls downwards because of the gravity of the earth.
- Gravitation may be the attraction of objects in outer space. Example: Force of attraction between the other planets and sun.

Each body in this universe attracts other bodies towards itself with a force known as Gravitational Force, thus gravitation is a study of the interaction between two masses. Out of the two masses, the heavier one is called source mass and the lighter one is called test mass.

Gravitational force is a central force which depends only on the position of test mass from the source mass and always acts along the line joining the centres of the two masses.

$$\vec{F}(r) = f(r)\hat{r}$$

The core problem of gravitation has always been in understanding the interaction between the two masses and the relativistic effects associated with it.

Newton's Law of Gravitation

According to Newton's law of gravitation, "Every particle in the universe attracts every other particle with a force whose magnitude is:

- Directly proportional to the product of their masses i.e. $F \propto (M_1M_2)$.
- Inversely proportional to the square of the distance between their centre i.e. $(F \propto 1/r^2)$.

On combining equations above we get,

$$F \propto M_1M_2 / r^2$$
$$F = G[M_1M_2]/r^2$$

Or, $f(r) = GM_1M_2/r^2$ [f(r)is a variable, Non-contact, and conservative force]. As f(r) varies inversely as a square of 'r' it is also known as inverse square law force. The proportionality constant (G) in the above equation is known as gravitational constant. The dimension formula of G is $[M^{-1}L^3T^{-2}]$. Also, the value of the gravitational constant:

- In SI units: 6.67×10^{-11} Nm² kg⁻².
- In CGS units: 6.67×10^{-8} dyne cm² g⁻².

Vector Form of Newton's Law of Gravitation

The vector form of Newton's law of gravitation signifies that the gravitational forces acting between the two particles form action-reaction pair.

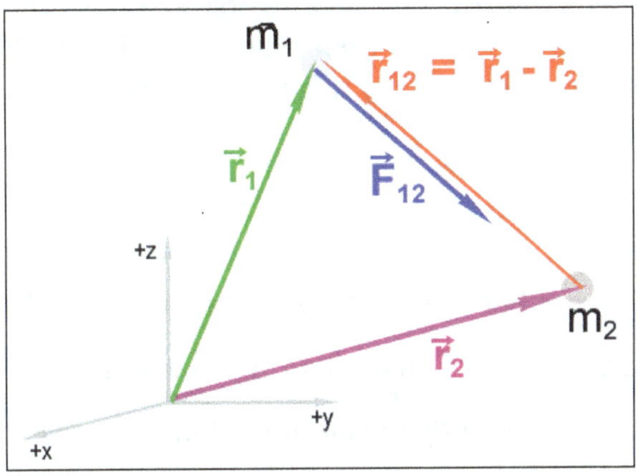

Newton's Law of Gravitation in Vector Form.

From the above figure, it can be seen that the two particles of masses and are placed at a distance, therefore according to Newton's law of gravitation, the force on m_1 due to m_2 i.e. F_{12} is given by,

$$F_{12} = [-G\, m_1 m_2]/|r_{12}|^2\, \hat{r}_{12}$$
$$F_{12} = [-Gm_1m_2/|r_1 - r_2|^3](r_1 - r_2)$$

Where $\hat{r}_{12}$ is a unit vector pointing from m_2 to m_1.

The negative sign in Equation above indicates that the direction of force F_{12} is opposite to that of $\hat{r}_{12}$.

Similarly, the force on m_2 due to m_1 i.e. F_{21} is given by,

$$F_{21} = -Gm_1m_2 / r_{21}|^2\, \hat{r}_{12}$$

From equations ($F_{12} = [-Gm_1m_2 / r_1 - r_2|^3](r_1 - r_2)$) and ($F_{21} = -Gm_1m_2 / r_{21}|^2\, \hat{r}_{12}$) we get,

$$F_{12} = -F_{21}$$

As F_{12} and F_{21} are directed towards the centres of the two particles, so the gravitational force is conservative in nature.

Gravitational Force Formula

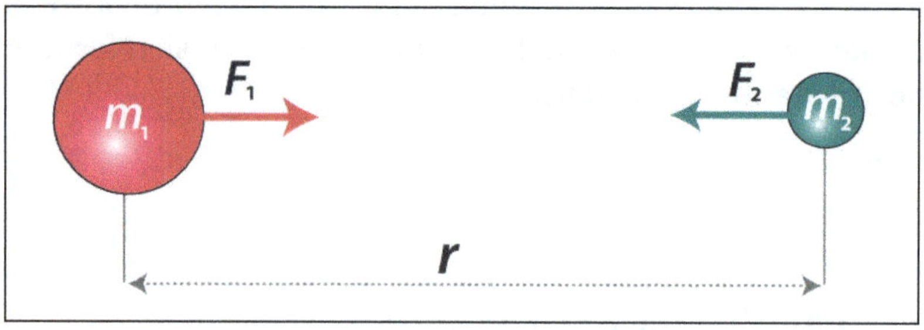

Gravitational force is explained using Newton's law of gravitation. Gravitational force decides how much we weigh and how far a ball travels when thrown before it lands on the ground. According to Newton's law of gravitation, every particle in the universe attracts every other particle with a force that is directly proportional to the product of their masses and inversely proportional to the square of the distance between them. Mathematically it can be represented as,

$$F = Gm_1m_2 / r^2$$

Where,

- F is the Gravitational force between two objects measured in Newton (N).
- G is the Universal Gravitational Constant with a value of 6.674×10^{-11} Nm^2kg^{-2}.
- m_1 is the mass of one massive body measured in kg.
- m_2 is the mass of another massive body measured in kg.
- r is the separation between them measured in kilometre (Km).

Principle of Superposition of Gravitational Forces

Newton's law of gravitation answers only the interaction between two particles if the system contains 'n' particles there are $n(n - 1)/2$ such interactions.

According to the principle of superposition, if each of these interactions acts independently and uninfluenced by the other bodies, the results can be expressed as the vector summation of these interactions:

$$F = F_{12} + F_{13} + F_{14} \ldots\ldots + F_{1n}.$$

It states that: "The resultant gravitational force F acting on a particle due to the number of point masses is equal to the vector sum of forces exerted by the individual masses on the given particle".

Derivation of Newton's Law of Gravitation from Kepler's Law

Suppose a test mass is revolving around a source mass in a nearly circular orbit of radius 'r', with a constant angular speed (ω). The centripetal force acting on the test mass for its circular motion is,

$$F = mr\omega^2 = mr \times (2\pi/T)^2$$

According to Kepler's 3rd law, $T^2 \propto r^3$. Using this in force equation we get,

$$F = 4\pi^2 mr/Kr^3 \text{ [Where, } K = 4\pi^2/GM]$$

$\Rightarrow$ F = GMm/r2, which is the equation of Newton's law of gravitation.

Light

Light is a form of electromagnetic radiation emitted by hot objects like lasers, bulbs, and the sun. Light is a kind of kinetic energy which enables the human eye to see or make things visible. Light energy is a form of electromagnetic radiation that can be seen by the human eye. Light contains photons which are minute packets of energy. When an object's atoms get heated up, it results in the production of photons and this is how photons are produced. The electrons find excitement from the heat and result in earning extra energy. The energy is released in the form of a photon and more photons come out as the substance gets hotter.

Waves and the Spectrum of Light

Light has the properties of waves. Like ocean waves, light waves have crests and troughs. The distance between one crest and the next, which is the same as the distance between one trough and the next, is called the wavelength. The frequency of a wave is the number of crests (or troughs) that pass a point in one second. The wavelength multiplied by the frequency equals the speed at which the wave travels.

The colours of visible light are red, orange, yellow, green, blue, indigo, and violet. These different colours of light have different wavelengths and frequencies. Red light has the longest wavelength, and the lowest frequency of the visible spectrum. Violet has the shortest wavelength, and the highest frequency of the visible spectrum. Look at the two waves in the picture below. You can imagine how, if they were both moving to the right at the same speed, the number of violet crests passing the edge of the box in one second would be higher than the number of red crests.

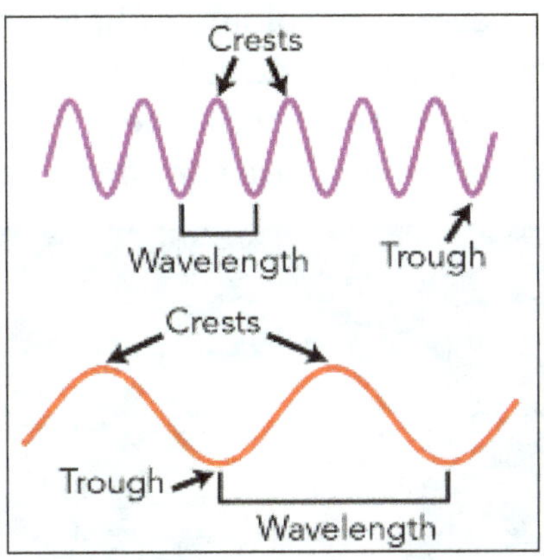

Red and violet light waves.

There is also light that is not visible to humans. Ultraviolet light and x-rays are also light, but have too small a wavelength and too high a frequency to be visible to us. Infrared light which can be detected by night-vision goggles, and radio waves, which are picked up by your radio so you can hear music, have wavelengths which are too long and frequencies which are too low to be seen by the human eye.

Chrysanthemum flower as seen using visible light (top),
ultraviolet light (middle) and infrared light (bottom).

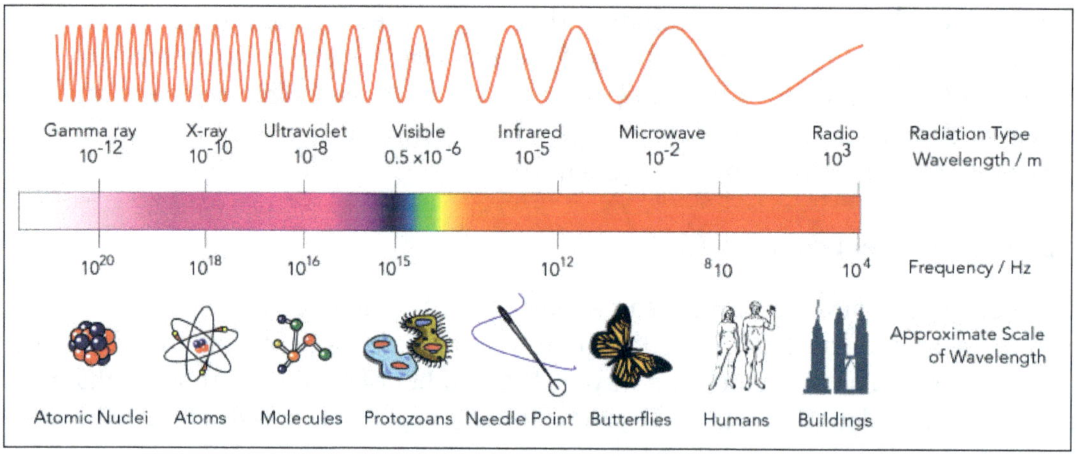

The electromagnetic spectrum.

Primary Colours of Light

You will remember from art class that the primary colours are red, yellow and blue. These can mix to form the secondary colours orange, green and purple. Light has primary colours as well. But these are different colours than the colours we use in paint and markers. The primary colours of light are red, green, and blue. The secondary colours of light are cyan (made by combining blue and green), magenta (made by combining blue and red) and yellow (made by combining green and red). Computer screens use various amounts of red, blue, and green light to make all the colours that you see. When the primary colours of light are combined, they make white light.

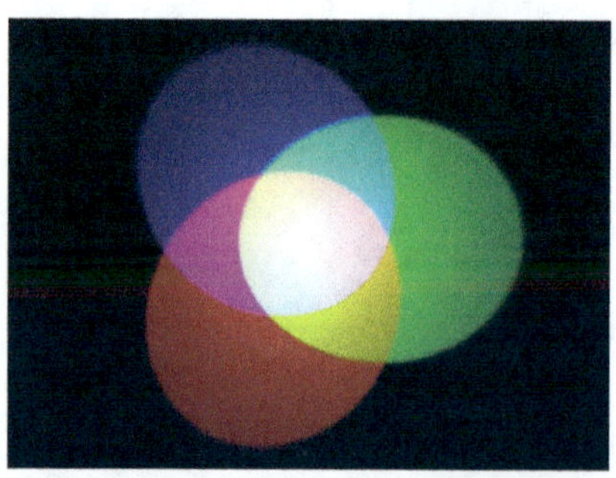

Primary colors of light. Blue, green, and red lights are shone on a black wall to show the secondary colours, with white light in the middle.

The human eye perceives colour using three types of photoreceptor cells which are sensitive to long, medium, and short wavelengths of visible light. Yellow wavelengths of light, for example, are perceived the same as a combination of red and green light, as in the image above. This is because they stimulate the cells in the eye in the same way. In other words, pure yellow light is physically different from a combination of red and green light, but they are both perceived by us as yellow. Did you know that two shades of green are easier for the human eye to differentiate, than the other colours? If you go into a paint shop and lay down all the red and all the green paint options, you will be able to more easily differentiate between the green shades than red. This is due to the fact that green is in the middle of the visible spectrum.

Wave

A wave is a disturbance in a medium that carries energy without a net movement of particles. It may take the form of elastic deformation, a variation of pressure, electric or magnetic intensity, electric potential, or temperature.

- Transfers energy.
- Usually involves a periodic, repetitive Movement.
- Does not result in a net movement of the medium or particles in the medium (mechanical wave).

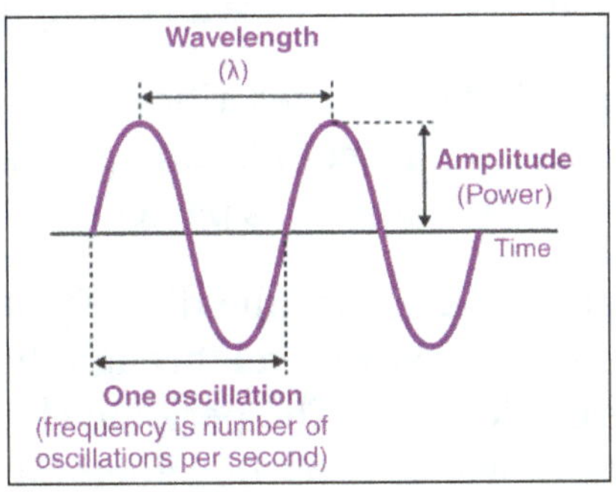

There are some basic descriptors of a wave. Wavelength is the distance between an identical part of the wave. Amplitude is the maximum displacement from the neutral position. This represents the energy of the wave. Greater amplitude carries greater energy.

Displacement is the position of a particular point in the medium as it moves as the wave passes. Maximum displacement is the amplitude of the wave Frequency (f) is the number of repetitions per second in Hz, s -1 Period (T) is the time for one wavelength to pass a point. T = f-1. The velocity (v) of the wave is the speed that a specific part of the wave passes a point. The speed of a light wave is c.

Types of Waves

The types of waves are given below:

- Transverse Waves: Waves in which the medium moves at right angles to the direction of the wave. Examples of transverse waves:
    - Water waves (ripples of gravity waves, not sound through water)

- Light waves
- S-wave earthquake waves
- Stringed instruments
- Torsion wave

The high point of a transverse wave is a crest. The low part is a trough.

- Longitudinal Wave: A longitudinal wave has the movement of the particles in the medium in the same dimension as the direction of movement of the wave. Parts of longitudinal waves are (1) Compression where the particles are close together. (2) Rarefaction: where the particles are spread apart. Examples of longitudinal waves:
  - Sound waves
  - P-type earthquake waves
  - Compression wave
- Electromagnetic Waves: Radio signals, light rays, x-rays, and cosmic rays.
- Mechanical waves: A wave which needs a medium in order to propagate itself. Sound waves, waves in a Slinky, and water waves are all examples of this.
- Matter Waves: Any moving object can be described as a wave When a stone is dropped into a pond, the water is disturbed from its equilibrium positions as the wave passes; it returns to its equilibrium position after the wave has passed.
- Electromagnetic Waves: These waves are the disturbance that does not need any object medium for propagation and can easily travel through the vacuum. They are produced due to various magnetic and electric fields. The periodic changes that take place in magnetic electric fields and therefore known as Electromagnetic Wave.

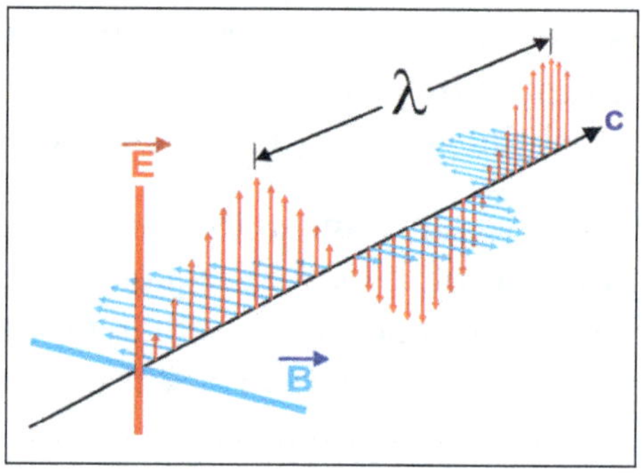

Wave Speed Formula

It is the total distance covered by the wave in a given time period. The formula for wave speed is given as,

Wave Speed = Distance Covered/Time taken

Properties of Waves

The prime properties of waves are as follows:

- Amplitude: Wave is an energy transport phenomenon. Amplitude is the height of the wave, usually measured in meters. It is directly related to the amount of energy carried by a wave.
- Wavelength: The distance between identical points in the adjacent cycles of crests of a wave is called a wavelength. It is also measured in meters.
- Period: The period of a wave is the time for a particle on a medium to make one complete vibrational cycle. As the period is time, hence is measured in units of time such as seconds or minutes.
- Frequency: Frequency of a wave is the number of waves passing a point in a certain time. The unit of frequency is hertz (Hz) which is equal to one wave per second.

  The period is the reciprocal of the frequency and vice versa.

$$\text{Period} = \frac{1}{\text{Frequency}}$$

Or,

$$\text{Frequency} = \frac{1}{\text{Period}}$$

- Speed: The speed of an object means how fast an object moves and is usually expressed as the distance travelled per time of travel. The speed of a wave refers to the distance travelled by a given point on the wave (crest) in a given interval of time. That is:

$$\text{Speed} = \frac{\text{Distance}}{\text{Time}}$$

  Speed of a wave is thus measured in meter/second i.e. m/s.

Relativity

Relativity is a theorem, formulated by Albert Einstein, which states that space and time are relative and all the motion must be relative to a frame of reference. It is a notion that states, laws of physics are the same everywhere. This theory is simple but hard to understand. It states:

- There is no absolute reference frame. One can measure velocity if the object or momentum is only in relation to other objects.
- The speed of light is constant irrespective of who measures it or how fast the person measuring it, is moving.

Albert Einstein's Theory of Relativity encompasses two theories, namely Special Relativity Theory and General Relativity Theory.

Special Theory of Relativity

Einstein first introduced this term in the year 1905. It is a theorem that deals with the structure of space-time. Einstein explained this theory based on two postulates:

- The laws of physics are the same for all irrespective of the velocity of the observer.
- The speed of light is always constant regardless of the motion of the light source or the motion of the observer.

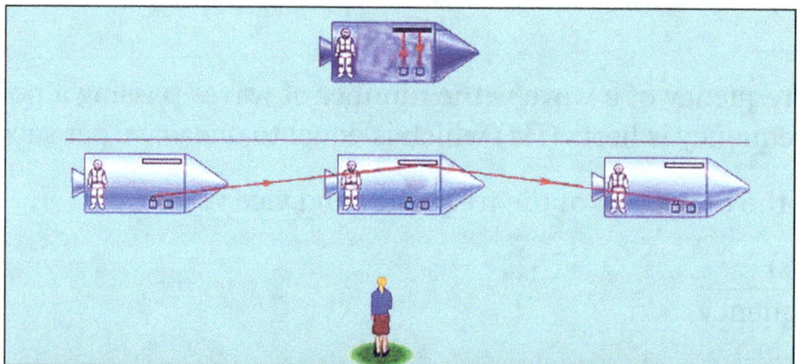

Earth observer sees light travel farther than does the astronaut

This is the theory which laid the foundation of time travel. According to Einstein, the rate at which time tics decreases with the increase of the velocity of the person. But this is hard to notice as the decrease in the time is relatively very low when compared to the increase in time. So, it can be assumed that if you can equal the velocity of light, you will be in a situation where time is still. This phenomenon is called Time Dilation. There are other surprising consequences of this theory such as:

- Relativity of simultaneity: Two actions, simultaneous for one person may not be simultaneous for another person in relative motion.
- Length Shrinking: Objects are measured and appeared to be shorter in the direction that they are moving with respect to the observer.
- Mass-Energy Equivalence: Study of relativity lead to one of the greatest inventions i.e., $E = mc^2$ where E is Energy, m stands for mass and c for the velocity of light. Many scientists observed that the mass of the object is increased with the velocity but never knew how to calculate it. This equation is the answer to their problem, which explains that the increased relativistic weight of the object is equal to the kinetic energy divided by the square of the speed of light.

General Theory of Relativity

General Relativity theory developed by Einstein in the year 1907-1915 states that being at rest in the gravitational field and accelerating are identical physically. For example, an observer can see the ball fall the same way on the rocket and on Earth. This is due to the acceleration of the rocket, which is equal to 9.8 m/s^2. This theory relates to Newton's gravitational theory and special relativity.

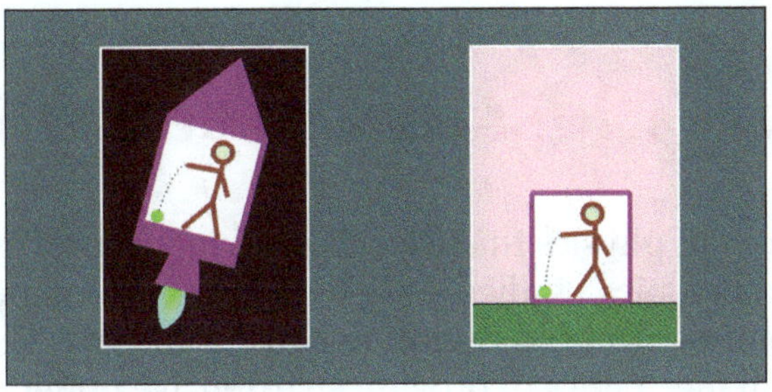

Some Consequences of General Relativity are:

- Gravitational Time Dilation: Gravity influences the passage of time. Clocks in the deeper gravitational wells run slower than in general gravitational levels.
- Light rays will bend in the gravitational field.
- The universe is expanding, and parts of it are moving away from Earth faster than the speed of the light.

References

- Momentum: britannica.com, Retrieved 11, August 2020
- Motion-in-physics: byjus.com, Retrieved 19, May 2020
- Force-physics: britannica.com, Retrieved 23, January 2020
- Energy-definition-and-examples-2698976: thoughtco.com, Retrieved 09, July 2020
- What-is-the-definition-of-light: byjus.com, Retrieved 17, April 2020
- Light-and-its-properties: letstalkscience.ca, Retrieved 28, June 2020
- Relativity: byjus.com, Retrieved 16, February 2020

2

Understanding Classical Mechanics

Classical mechanics is a theory of physics that explains the motion of objects under the influence of forces. Some of the important topics studied under classical mechanics are Hamilton's principle and Lagrange's equation, rotating frame and motion of a rigid body. This chapter closely examines these key concepts of classical mechanics to provide an extensive understanding of the subject.

Classical mechanics is used for describing the motion of macroscopic objects, from projectiles to parts of machinery, as well as astronomical objects, such as spacecraft, planets, stars, and galaxies. It produces very accurate results within these domains, and is one of the oldest and largest subjects in science, engineering and technology.

Besides this, many related specialties exist, dealing with gases, liquids, and solids, and so on. Classical mechanics is enhanced by special relativity for objects moving with high velocity, approaching the speed of light; general relativity is employed to handle gravitation at a deeper level; and quantum mechanics handles the wave-particle duality of atoms and molecules.

In physics, classical mechanics is one of the two major sub-fields of study in the science of mechanics, which is concerned with the set of physical laws governing and mathematically describing the motions of bodies and aggregates of bodies. The other sub-field is quantum mechanics.

Hamilton's Principle and Lagrange's Equation

Hamilton's principle is a variational principle from which can be derived the equations of motion of a classical dynamical system in which friction or other forms of dissipation of energy do not occur. In the original formulation of Newton's laws of motion, the position of each particle of the system of interest is specified by the cartesian coordinates of that particle. In many cases, these coordinates are not all independent of each other or do not reflect the structure of the system in a convenient way. It is then advantageous to introduce a system of generalized coordinates which are independent of each other and do reflect any special features of the system such as its symmetry about some center. The number of degrees of freedom of the system, f, is the number of such coordinates required to specify the configuration of the system at any time.

The general principle we need was given in its final form by Hamilton, and is often called the principle of least action. The term "action" refers to the integral of L over time:

$$J = \int_{t_1}^{t_2} L(q, \dot{q}, t)\,dt$$

Here the limits are two times at which the particle has two different states. We imagine that these times and the corresponding states are fixed, but that we can vary both q and $\dot{q}$ during the time in between, making the particle follow different paths, so that J is varied. Calling these variations δq, $\delta \dot{q}$, and δJ, we have,

$$\delta J = \int_{t_1}^{t_2} L(q+\delta q, \dot{q}+\delta\dot{q}, t)dt - \int_{t_1}^{t_2} L(q, \dot{q}, t)dt$$

Hamilton's principle says that for the actual motion of the particle, δJ = 0 to first order in the variations δq and $\delta \dot{q}$. That is, the actual motion of the particle is such that small variations do not change the action.

Now by Taylor's theorem we can write to 1st order,

$$L(q+\delta q, \dot{q}+\delta\dot{q}, t)dt \approx L(q, \dot{q}, t) + \frac{\partial L}{\partial q}\delta q + \frac{\partial L}{\partial \dot{q}}\delta\dot{q}$$

where the partial derivatives are evaluated for $\delta q = \delta \dot{q} = 0$. Thus we find,

$$\delta J = \int_{t_1}^{t_2} \left[\frac{\partial L}{\partial q}\delta q + \frac{\partial L}{\partial \dot{q}}\delta\dot{q} \right] dt$$

Since $\dot{q} = dq/dt$ we have $\delta \dot{q} = d(\delta q)/dt$, so the 2nd term in the integral is,

$$\int_{t_1}^{t_2} \frac{\partial L}{\partial \dot{q}} \frac{d(\delta q)}{dt} dt$$

and we integrate by parts to convert this to,

$$\frac{\partial L}{\partial \dot{q}}\delta q \Big|_{t_1}^{t_2} - \int_{t_1}^{t_2} \frac{d}{dt}\left(\frac{\partial L}{\partial \dot{q}}\right) \cdot \delta q \ dt.$$

Because the states at the initial and final times are fixed, δq vanishes at both times, so the first term above is zero. We have then,

$$\delta J = \int_{t_1}^{t_2} \left[\frac{\partial L}{\partial q} - \frac{d}{dt}\left(\frac{\partial L}{\partial \dot{q}}\right) \right] \delta\dot{q} \ dt = 0$$

Since δq is arbitrary, the quantity in [ ] must vanish, so we have finally,

$$\frac{d}{dt}\left(\frac{\partial L}{\partial \dot{q}}\right) - \frac{\partial L}{\partial q} = 0$$

It becomes a differential equation (2nd order in time) to be solved. It is the equation of motion for the particle, and is called Lagrange's equation. The function L is called the Lagrangian of the system.

Properties of the Lagrangian

So Hamilton's principle has given us Equation for the Lagrangian. What do we know about L beyond the variables it depends on? We assume we are in an inertial reference frame. Then all coordinate axes are equivalent, so L must be a scalar. And our choice of when to start the clocks is arbitrary, so L cannot depend explicitly on t.

Beyond that we can make some reasonable requirements. Suppose we have two systems A and B separated by large distances so they do not interact with each other. Then the Lagrangian for this composite system must consist of separate parts for each, i.e., L(A + B) = L(A) + L(B) . Furthermore, multiplying L by some constant would change nothing in the equations so far. Choice of that constant simply involves choosing a system of units.

Another thing that does not change the physical content of the Lagrangian is adding to it the total time derivative of a function of q and t. Suppose we define,

$$L'(q,\dot{q},t)=L(q,\dot{q},t)+\frac{df(q,t)}{dt}$$

Then since,

$$\int_{t_1}^{t_2}\frac{df}{dt}dt=f(t_2)-f(t_1)$$

for the action we have,

$$J'=J+f(t_2)-f(t_1)$$

The terms in f evaluated at the end points do not change when we perform the variation, so δJ' =δJ. The two Lagrangians give the same variation and are thus equivalent in physical content.

Now we take the simplest system, a particle moving without any interaction with the external world. We know its velocity is constant (the 1st law). Since all points in space are equivalent for such a particle, L cannot depend on its position x. It must therefore depend only on the velocity v. But it is a scalar, so it can depend only on v^2. We have that ! $\partial L/\partial x_i = 0$, so by Lagrange's equation,

$$\frac{d}{dt}\left(\frac{\partial L}{\partial v_i}\right)=0$$

showing that $\partial L/\partial v_i$ is constant. But,

$$\frac{\partial L}{\partial v_i}=\frac{\partial L}{\partial v^2}\cdot\frac{\partial v^2}{\partial v_i}=2\frac{\partial L}{\partial v^2}\cdot v_i$$

and we know that v_i is constant. This means,

$$\frac{\partial L}{\partial v^2}=\text{const}$$

We conclude that L = (const) · v^2. We choose the constant to be 1/2m and have,

$$L=\frac{1}{2}mv^2$$

The kinetic energy T of the particle.

Now we introduce interactions of the particle with its environment, In Newtonian mechanics these are described by forces, the connection to the motion being given by the 2nd law. We try to introduce these into to the Lagrangian by adding a term to the one we already have.

Suppose the interaction term in L does not depend explicitly on the particle's velocity. Then we will have $\partial L/\partial v_i = \partial T/\partial v_i = mv_i$, and Lagrange's equation becomes,

$$\frac{d}{dt}(mv_i) - \frac{\partial L}{\partial x_i} = 0$$

or

$$m\ddot{x}_i \frac{\partial L}{\partial x_i}.$$

For this to give us the 2nd law we need the right side to be the force. We know this to be given by $-\partial U(x)/\partial x_i$, where U is the potential energy function for the force. We are thus led to the final form for the Lagrangian:

$$L(x_i, v_i, t) = T - U(x_i, t)$$

The possible dependence of U on t might arise if the locations of objects with which our particle interacts are changing with time in a known way. In most of our cases U will not depend on t.

Advantages of the Lagrange Formulation

Perhaps the main advantage of the Lagrange approach is its use of generalized coordinates. This allows use of different coordinates for different parts of the system. Example: A small block of mass m starts from rest at the top of a frictionless wedge of mass M which is on a frictionless horizontal floor. The block slides down the wedge, while the wedge slides across the floor. We wish to find the equations of motion for the block and the wedge. Choose s to represent the distance the block moves down the wedge, and x to represent the distance the wedge moves across the floor. These are our generalized coordinates.

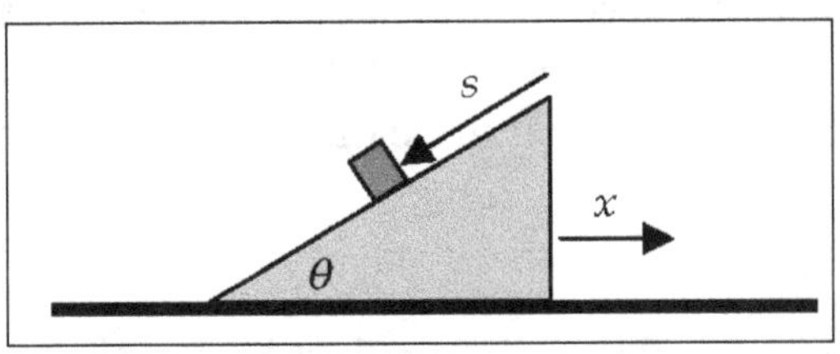

First we construct the total kinetic energy, T. The velocity of the wedge is $\dot{x}$. We break the block's velocity (in the inertial frame of the floor) into vertical and horizontal components: $v_x = \dot{x} - \dot{s}\cos\theta$, $v_y = -\dot{s}\sin\theta$. So for the combined system we have,

$$T = \frac{1}{2}M\dot{x}^2 + \frac{1}{2}m\left(\dot{x}^2 + \dot{s}^2 - 2\dot{x}\dot{s}\cos\theta\right)$$

The potential energy (taking U = o at the top of the wedge) is,

$$U = -mgs\ \sin\theta.$$

Therefore the Lagrangian is,

$$L = \frac{1}{2}m\dot{x}^2 + \frac{1}{2}m\left(\dot{x}^2 + \dot{s}^2 - 2\dot{x}\dot{s}\cos\theta\right) + mgs\ \sin\theta$$

We have $\frac{\partial L}{\partial x} = o$, $\frac{\partial L}{\partial s} = mg\ \sin\theta$, $\frac{\partial L}{\partial \dot{x}} = (M+m)\dot{x} - m\dot{s}\ \cos\theta$, $\frac{\partial L}{\partial \dot{s}} = m\dot{s}\cos\theta$. The Lagrange equations are therefore,

$$(M+m)\ddot{x} - m\ddot{s}\ \cos\theta = o$$
$$m\ddot{s} - m\ddot{x}\cos\theta - mg\sin\theta = o$$

Solving the first one for $\ddot{x}$ and substituting in the second one, we find after some rearrangement,

$$\ddot{s} = \frac{M+m}{M+m\sin^2\theta}.g\sin\theta$$

This is a constant acceleration, so it is easy to find s at any time. Note that as $M \to \infty$ this gives $\ddot{s}$ = g sinθ, as it should, since the wedge will not move. For the acceleration of the wedge we find, using the first Lagrange equation above,

$$\ddot{x} = \frac{m}{M+m\sin^2\theta}.g\sin\theta\ \cos\theta$$

This is also constant. To solve this problem using the 2nd law we would need to bring in the normal force between the wedge and the block, which we do not know. To use the free body diagram for the block on the wedge we would have to take into account that the wedge is accelerating, so this is a non-inertial frame. The Lagrange method is much easier.

In this example there is a constraint on the motion of the block: it must move along the incline of the wedge. We built that in when we wrote the components of the block's velocity, using ($-\dot{s}$ cosθ,$-\dot{s}$ sinθ) for the components of the vector $\dot{s}$. One can approach this problem by using as generalized coordinates the horizontal and vertical coordinates of the block plus the horizontal coordinate of the wedge, and then imposing separately the constraint that it must move along the incline. The best way to do this is through the use of what are called undetermined multipliers.

Conservation Laws

As the importance of energy became clearer in the first half of the 19th century the reformulation of classical mechanics took another turn, especially through the work of Hamilton. We consider a closed system, which means one where there are no interactions with anything outside the system. Different parts of the system can interact in complicated ways, but it is free of outside influences.

Presumably it occupies a finite region of space. So to observers inside the system the space outside is infinite in all directions and all directions are equivalent to it. That is, where one puts the origin of a coordinate system, and how one orients the axes, is totally arbitrary. Equally arbitrary is when one decides to start clocks to count time. We say that for such a system space and time are homogeneous.

We start with the consequence of homogeneity in time. This means that the Lagrangian L describing the system cannot depend explicitly on t. The time derivative of L is thus,

$$\frac{dL}{dt} = \frac{\partial L}{\partial \dot{q}_i}.\ddot{q}_i + \frac{\partial L}{\partial q_i}.\dot{q}_i$$

(Sum over i implied.) We use the Lagrange equations to substitute for $\frac{\partial L}{\partial q_i}$ and find,

$$\frac{dL}{dt} = \frac{\partial L}{\partial \dot{q}_i}.\ddot{q}_i + \dot{q}_i \frac{d}{dt}\left(\frac{\partial L}{\partial \dot{q}_i}\right) = \frac{d}{dt}\left[\dot{q}_i \frac{\partial L}{\partial q_i}\right]$$

We write this as,

$$\frac{d}{dt}\left[\dot{q}_i \frac{\partial L}{\partial \dot{q}_i} - L\right] = 0$$

The quantity in [ ] is thus constant in time. Since L = T - U, where T is a quadratic function of the $\dot{q}_i$ and U is independent of them, it follows that,

$$\dot{q}_i \frac{\partial L}{\partial \dot{q}_i} = 2T$$

This is true for any generalized coordinates, by a theorem of Euler, but it is easy to show in cartesian coordinates.

So the quantity in [ ] above is T + U, and what we have found is that this quantity is constant in time. This is the mechanical energy of the system, so we have proved that homogeneity in time implies that the total mechanical energy of a closed system is conserved. A remarkable conclusion, showing just how fundamental energy is.

A similar argument follows from the fact that for a closed system we can move every particle the same small distance without changing any behavior of the system. That is, the system's location in space is irrelevant because space is homogeneous.

We will use cartesian coordinates. Let the ath particle's position vector, x_i^a, be changed to $x_i^a + \varepsilon_i$, where ε_i is a small distance. This will change the Lagrangian by a small amount:

$$\delta L = L\left(x_i^a + \varepsilon_i\right) - L\left(x_i^a\right) = \sum_a \varepsilon_i \frac{\partial L}{\partial x_i^a} = \varepsilon_i \sum_a \frac{\partial L}{\partial x_i^a}$$

to first order in the ε's. But this movement of all the particles by the same amount cannot have any effect, so we require $\delta L = 0$. Since the ε's are arbitrary this implies,

$$\sum_a \frac{\partial L}{\partial x_i^a} = 0$$

Now we use the Lagrange equations $\frac{\partial L}{\partial x_i^a} = \frac{d}{dt}\left(\frac{\partial L}{\partial v_i^a}\right)$, where $v_i^a = x_i^a$ is the, corresponding velocity component, to find,

$$\sum_a \frac{d}{dt}\left(\frac{\partial L}{\partial v_i^a}\right) = \frac{d}{dt}\sum_a\left(\frac{\partial L}{\partial v_i^a}\right) = 0$$

So the sum is constant in time. What physical quantity is it? Since the potential energy does not depend on the velocities, only the kinetic energy is involved, and it can be written as,

$$T = \sum_a \frac{1}{2} m_a \left(v_i^a v_i^a\right)$$

Thus we have,

$$\frac{\partial L}{\partial v_i^a} = m_a v_i^a = p_i^a$$

the ith component of the momentum of the ath particle. So what we have proved is that,

$$\frac{d}{dt}\sum_a p_i^a = 0.$$

The sum is the total linear momentum (ith component) of the system. We have shown that the homogeneity of space implies conservation of the total linear momentum of a closed system. Another remarkable result.

The conservation law holds for each component of momentum separately. That is, if the Lagrangian remains unchanged when the system is moved in a particular direction, the total momentum component in that direction is conserved, whether it is true for other directions or not.

A similar argument, shows that if rotation of the whole system about some axis does not change the Lagrangian (as it will not for a closed system) then the total angular momentum about that axis is conserved. And again, this holds regardless of whether it is valid about other axes.

The lesson learned is that the conservation laws are direct consequences of symmetries of the system. In Newtonian mechanics those symmetries take the form of statements about external force and/or torques being zero. Here they are simple statements about the Lagrangian.

Hamilton's Formulation

We saw in here in discussion of conservation of linear momentum that the linear momentum of a particle is given by,

$$p_i = \frac{\partial L}{\partial v_i} = \frac{\partial L}{\partial \dot{x}_i}$$

We were using cartesian coordinates, of course. But the same kind of relation holds when we use generalized coordinates. What we get is called the generalized momentum:

$$p_i = \frac{\partial L}{\partial \dot{q}_i}$$

It may not have the dimensions of mass times velocity if q_i is not a length, but that doesn't matter. In Hamilton's reformulation the variables describing the state of a particle are taken to be the set (q_i, p_i) rather than the set $(q_i, \dot{q}_i)$.

We return to Equation ($\frac{d}{dt}\left[\dot{q}_i \frac{\partial L}{\partial \dot{q}_i} - L\right] = 0$) above and write it as,

$$\frac{d}{dt}(\dot{q}_i p_i - L) = 0$$

where L is now regarded as function of the set (q_i, p_i). The function in ( ) above is called the Hamiltonian of the particle:

$$H(\dot{q}_i, p_i) = \dot{q}_i p_i - L$$

For a system, one sums the first term over all the particles and uses the proper L for the whole system. We have seen that if L does not depend explicitly on time, the Hamiltonian is the same as the total energy E = T + U, and it is conserved. But the function is useful even if it does depend explicitly on t and is not the conserved total energy. The total differential of H is,

$$dH = \frac{\partial H}{\partial q_i} dq_i + \frac{\partial H}{\partial p_i} dp_i + \frac{\partial H}{\partial t} dt$$

but by Equation ($H(\dot{q}_i, p_i) = \dot{q}_i p_i - L$) it is also,

$$dH = \dot{q}_i dp_i + p_i d\dot{q}_i - \frac{\partial L}{\partial q_i} dq_i - \frac{\partial L}{\partial \dot{q}_i} d\dot{q}_i - \frac{\partial L}{\partial t} dt$$

In this expression the 2nd and 4th terms on the right are the same, so they cancel. Now from the Lagrange equations,

$$\frac{\partial L}{\partial q_i} = \frac{d}{dt}\left(\frac{\partial L}{\partial \dot{q}_i}\right) = \frac{d}{dt}(p_i) = \dot{p}_i$$

so Equation ($dH = \dot{q}_i dp_i + p_i d\dot{q}_i - \frac{\partial L}{\partial q_i} dq_i - \frac{\partial L}{\partial \dot{q}_i} d\dot{q}_i - \frac{\partial L}{\partial t} dt$) becomes,

$$dH = \dot{q}_i dp_i - \dot{p}_i dq_i - \frac{\partial L}{\partial t} dt$$

Comparing with Equation ($dH = \frac{\partial H}{\partial q_i} dq_i + \frac{\partial H}{\partial p_i} dp_i + \frac{\partial H}{\partial t} dt$) we find,

$$\dot{q}_i = \frac{\partial H}{\partial p_i}$$

$$\dot{p}_i = \frac{\partial H}{\partial \dot{q}_i}$$

which are called Hamiltion's equations of motion.

We also find $\frac{\partial H}{\partial t} = -\frac{\partial L}{\partial t}$, which is of interest only for systems depending explicitly on time. Extending our earlier analysis to include such a case, we find $\frac{dH}{dt} = \frac{\partial H}{\partial t}$. That is, H is conserved unless it depends explicitly on time.

The content of Hamilton's equations can be made clear by looking at a single particle subject to a conservative force with potential energy function $U(q_i)$. The kinetic energy is,

$$T = \frac{1}{2} m v_i v_i = \frac{p_i p_i}{2m}$$

so we have,

$$H(q_i p_i) = \frac{p_i p_i}{2m} + U(q_i)$$

Hamilton's equations then give us,

$$\dot{q}_i = \frac{p_i}{m} \text{ and } \dot{p}_i = -\frac{\partial U}{\partial q_i}$$

The first is the definition of momentum, the second is Newton's 2nd law. Turning to generalized coordinates, the second Hamilton equation tells us that if some variable q_i is missing from the Hamiltonian, then the corresponding momentum is conserved. Such variables are called cyclic. For example, if, in polar coordinates, we find that H does not depend explicitly on the azimuthal angle φ, then the angular momentum along the polar axis will be conserved.

Hamilton's formulation is of course logically equivalent to Lagrange's and Newton's. Its two first order (in time) differential equations are mathematically equivalent to the second order Lagrange equations. But the effects of the symmetry of the situation are often much easier to find and make use of in the Hamiltonian version. And, it turns out, the transition from the classical approximation — which is what this is — to the more general theory, quantum mechanics is easier to make if one starts from the Hamiltonian formulation.

Rotating Frame

Newton's Law of Motion are valid in an inertial frame of reference which can be chosen to be the frame with 'fixed stars'. The earth is in complex motion with respect to the fixed stars. It is enough to treat the coordinate system fixed on earth to be undergoing rotations about its axis. We thus have to deal with non-inertial rotating frame of reference.

Rotating Coordinate System

We will consider two sets of coordinate axis. One set of coordinate is fixed in an 'inertial frame' Σ_{In} and Σ_{rot} other in a frame which is in arbitrary motion with respect to the inertial frame. We will take the non-inertial frame to be a rotating frame. The coordinates in the inertial or in fixed frame will be designated as x_i' or $(x_i)_{fixed}$ and in the rotating frame as x_i or $(x_i)_{rot}$. Let us consider a point P whose position vector is given by the vector in the rotating frame. Let be the position vector of the origin of the rotating coordinate system with respect to the fixed inertial frame. The position vector of P in the inertial frame is now given by,

$$r' = R + r$$

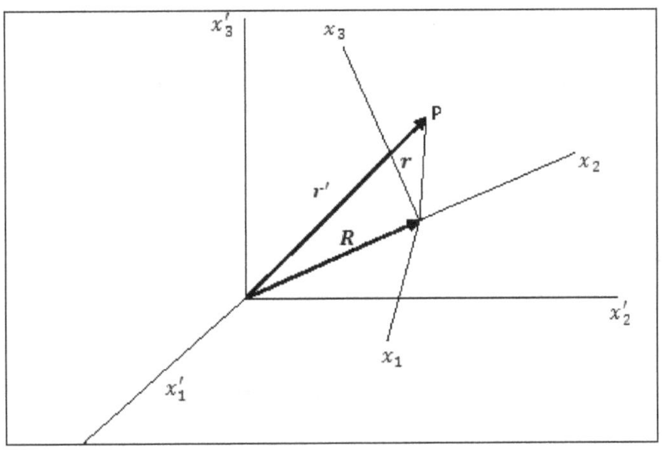

Now if the frame Σ_{rot} undergoes an infinitesimal rotation dθ, the displacement of P in the fixed frame Σ_{In} will be described by,

$$(dr)_{fixed} = d\theta \times r$$

We can do so because as we saw earlier an arbitrary infinitesimal displacement can always be represented by a pure rotation about some axis. In equation above the infinitesimal displacement dr is measured in the fixed inertial coordinate system x'. The infinitesimal displacement takes place in time dt, we therefore, have,

$$\left(\frac{dr}{dt}\right)_{fixed} = \frac{d\theta}{dt} \times r$$

From ($(dr)_{fixed} = d\theta \times r$) and if the angular velocity of rotation is,

$$\omega = \frac{d\theta}{dt}$$

We have,

$$\left(\frac{dr}{dt}\right)_{fixed} = \omega \times r$$

Remember the point P is fixed in the frame Σ_{rot} and it is the coordinate system itself that undergoes rotation.

Now let the point P has a velocity in the rotating frame given by $\left(\frac{dr}{dt}\right)_{rot}$ with respect to x_i the coordinate system. Thus the rate of change of r in the fixed system is given by,

$$\left(\frac{dr}{dt}\right)_{fixed} = \left(\frac{dr}{dt}\right)_{rot} + \omega \times r$$

This result is valid for any arbitrary vector, any arbitrary vector being defined as a physical quantity that transforms according to the same rules as the position vector. Thus for any vector A

$$\left(\frac{dA}{dt}\right)_{fixed} = \left(\frac{dA}{dt}\right)_{rot} + \omega \times r$$

We next see that the angular acceleration is invariant under this transformation.

$$\left(\frac{d\omega}{dt}\right)_{fixed} = \left(\frac{d\omega}{dt}\right)_{rot} + \omega \times \omega$$

and since $\omega \times \omega$ vanishes,

$$(\dot{\omega})_{fixed} = (\dot{\omega})_{rot}$$

The velocity of a particle P measured in the fixed inertial coordinate system from (r' = R+r) is,

$$\left(\frac{dr'}{dt}\right)_{fixed} = \left(\frac{dR}{dt}\right)_{fixed} + \left(\frac{dr}{dt}\right)_{fixed}$$

Substituting for $\left(\frac{dr}{dt}\right)_{fixed}$ from ($\left(\frac{dr}{dt}\right)_{fixed} = \left(\frac{dr}{dt}\right)_{rot} + \omega \times r$) we get,

$$\left(\frac{dr'}{dt}\right)_{fixed} = \left(\frac{dR}{dt}\right)_{fixed} + \left(\frac{dr}{dt}\right)_{fixed} + \omega \times r$$

Or,

$$v_{fixed} = v + v_{rot} + \omega \times r$$

Where,

- v_{fixed} is the velocity w.r.t. the fixed axis.
- v is the velocity of the frame Σ_{rot} w.r.t. Σ_{In}.
- v_{rot} is the velocity in the rotating frame
- ω is the angular velocity of the rotating frame.
- ω x r is the velocity due the rotation of the rotating frame.

Coriolis Force

In classical mechanics, Coriolis force is also known as Coriolis effect, an inertial force described by the 19th-century French engineer-mathematician Gustave-Gaspard Coriolis in 1835. Coriolis showed that, if the ordinary Newtonian laws of motion of bodies are to be used in a rotating frame of reference, an inertial force—acting to the right of the direction of body motion for counterclockwise rotation of the reference frame or to the left for clockwise rotation—must be included in the equations of motion.

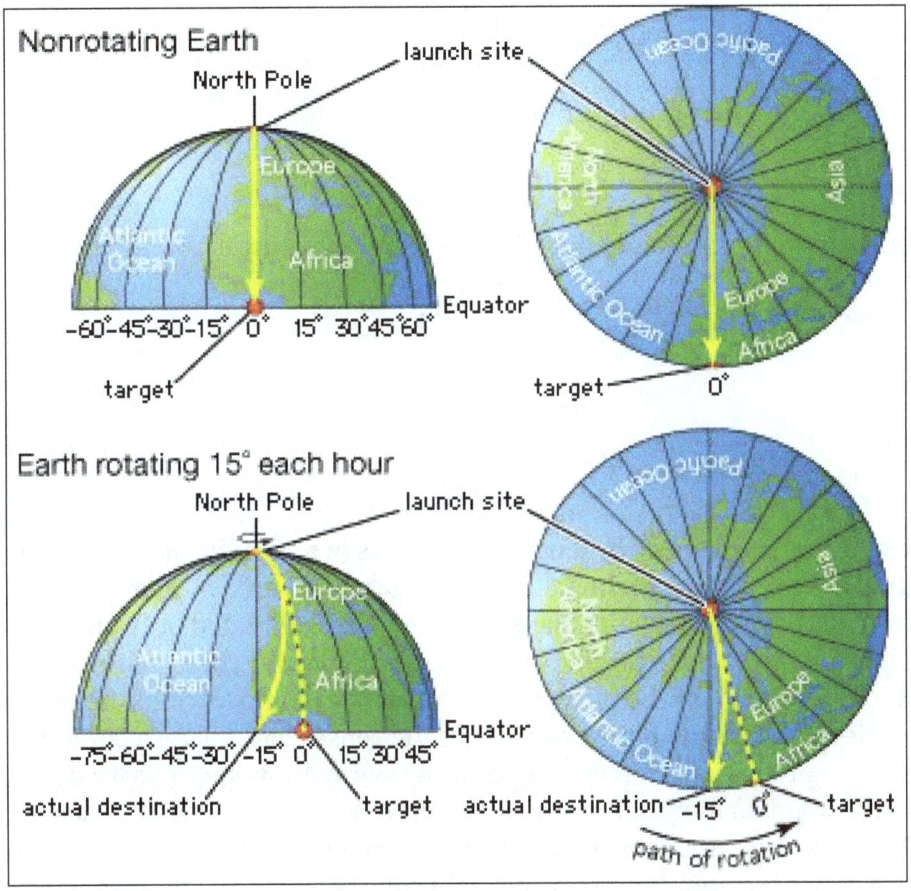

As seen from a fixed point in space, a parcel of air would be moving in a straight line. This apparent force on the motion of a fluid (in this case, air) is called the Coriolis effect. As a result of the Coriolis effect, air tends to rotate counterclockwise around large-scale low-pressure systems and clockwise around large-scale high-pressure systems in the Northern Hemisphere. In the Southern Hemisphere, the flow direction is reversed.

Force on a particle in the rotating frame can be obtained from the expression of Newton's equation of motion which is valid in the inertial (fixed) frame namely,

$$F_{inertial} = ma_{inertial} = m\left(\frac{dv}{dt}\right)_{fixed}$$

From equation $v_{fixed} = v + v_{rot} + \omega \times r$

$$\left(\frac{dv}{dt}\right)_{fixed} = \frac{dv}{dt} + \left(\frac{dv_{rot}}{dt}\right)_{fixed} + \omega \times \left(\frac{dr}{dt}\right)_{fixed}$$

We will first assume that the velocity of the rotating frame with respect to the inertial frame is constant so that $\frac{dv}{dt} = 0$. Substituting for $\left(\frac{dv_{rot}}{dt}\right)_{fixed}$ from equation ($\left(\frac{dA}{dt}\right)_{fixed} = \left(\frac{dA}{dt}\right)_{rot} + \omega \times r$) and for $\left(\frac{dr}{dt}\right)_{fixed}$ from equation ($\left(\frac{dr}{dt}\right)_{fixed} = \left(\frac{dr}{dt}\right)_{rot} + \omega \times r$) we get,

$$F_{inertial} = m\left(\frac{dv}{dt}\right)_{rot} + m\omega \times \omega \times r - 2m\omega \times v_{rot}$$

$$= m\,a_{inertial}$$

For an observer in the rotating frame, the effective force on a particle would be,

$$F_{eff} = m\,a_{rot} = m\,a_{inertial} - m\omega \times \omega \times r - 2m\omega \times v_{rot}$$

where,

$$a_{rot} = \left(\frac{dv}{dt}\right)_{rot}$$

Thus the effective force experienced by a particle consists of a term from Newton's Law, a term $-m\,\omega \times \omega \times r$ which is the familiar 'centrifugal force' term (which equals $-m\omega^2 r$) when ω is normal to the radius vector and the negative sign means that the force is directed outwards from the centre of rotation. In addition if the particle is moving in the rotating frame there exists another force which is proportional to the velocity of the particle. The 'centrifugal' and 'coriolis' forces are not forces in the usual sense. Nevertheless the forces are real in the sense that one feels an outward centrifugal force in a merry-go-round. These forces can be dispensed with if we prefer to describe the motion of a dynamical system in the inertial frame only. They arise because we have chosen to describe the motion using Newton's law in a 'non-inertial frame' in the present case a 'rotating frame'. Thus in order to write Newton's equation in a non-inertial frame we need to add non-inertial terms in the expression of the force along with the usual force F,

$$F_{eff} = F_{inertial} + \text{non-inertial terms}$$

The Coriolis force like the centrifugal force arises only in a rotating frame of reference. If for example we want to describe the motion of a particle under gravity on earth, in addition to the force of gravity acting on the particle, we will need to take into account the centrifugal and coriolis force. The coriolis force associated with earth's rotation is quite weak because the earth rotates about its axis only once in a day. This corresponds to an angular speed $\omega \simeq 2\pi \times 10^{-5}$ rad s^{-1}. If an object is moving with a speed of 1000 ms^{-1}, the maximum possible coriolis force on it is of the order of 10^{-2} ms^{-2} which is much smaller than $g \simeq 10\,ms^{-2}$. Coriolis force may be weak but it has important consequences. On earth it acts to change the direction of a moving body to the right in the Northern Hemisphere and to the left in the Southern Hemisphere.

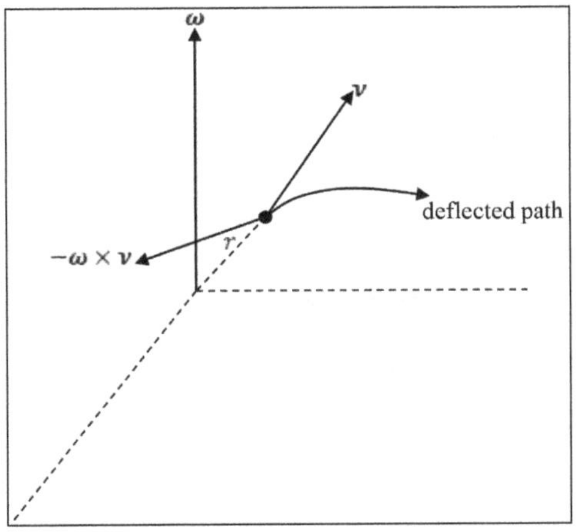

The centrifugal and coriolis forces are responsible for large scale atmospheric circulation, in the development of storms and in the sea-breeze circulation. It is important to incorporate coriolis force in large distance fights and in long-range ballistics.

Example: A gunner fired a shot in the Northern hemisphere at latitude of 45°. The gun is fixed at an angle of elevation α towards north. Show that if the range is 20 Km, the shell falls 44m towards east from the mark and if the gun was fired in the southern hemisphere, the shot would have fallen 88 m towards west. Set up the coordinates as shown,

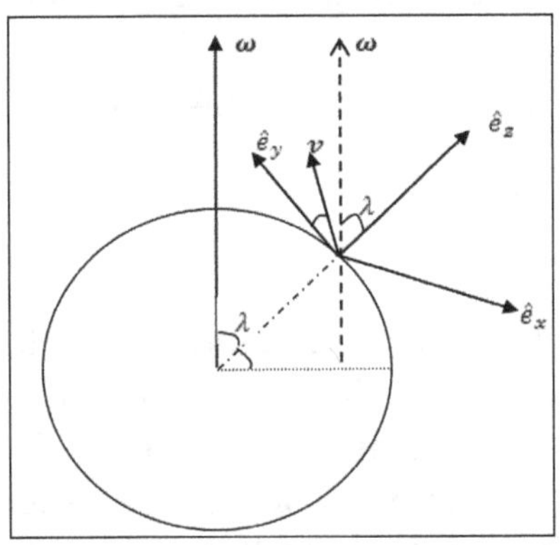

The z-axis is along the line joining the centre of the earth with the point P on the surface from where the gun is fired. Choose y and x axis as shown such that the gun is fired in the y-z plane. The angular velocity ω has components given as,

$$\omega = (0, \omega \sin\lambda, \omega \cos\lambda)$$

The velocity of the shot makes an angle λ with the y-axis and is in the y-z plane. Its components are,

$$v = (0, v \cos\alpha, v \sin\alpha - gt)$$

the acceleration along the z-axis is -g. The effect of centrifugal force which acts along the +ve z direction is to modify the acceleration due to gravity.

The equation of motion ($a_{rot} = \left(\frac{dv}{dt}\right)_{rot}$) gives,

$$\omega \times v = \begin{vmatrix} \hat{e}_x & \hat{e}_y & \hat{e}_z \\ 0 & \omega \sin\lambda & \omega \cos\lambda \\ 0 & v \cos\alpha & v \sin\alpha - gt \end{vmatrix}$$

giving,

$$m\ddot{x} = -2m v\omega\{\sin\lambda(\sin\alpha - gt) - \cos\lambda \cos\alpha\}$$

$$m\ddot{x} = -2m v\omega \cos(\alpha+\lambda) + 2m v\omega \sin\lambda t$$

$$m\ddot{z} = -mg$$

Equation above can be easily integrated, we get,

$$x = 2v\omega \cos(\alpha+\lambda)t^2 + \frac{1}{3} v\omega g \sin\lambda t^3$$

The time of flight,

$$t = 2v\frac{\sin\alpha}{g}$$

and the range is given by $d = \frac{v^2}{g}$. Substituting d = 20 x 10^3m, $\lambda = 45^o$ for the maximum range g = 10 ms^2, $\omega = 2\pi$ x 10^{-s} $rads^{-1}$. We get $x \simeq 44m$. If the gun was fired in southern hemisphere $\alpha = 45^o$ and $\lambda = 45^o, x \simeq -44m$.

The Foucault Pendulum

Foucault pendulum is a heavy metal sphere suspended by a long wire. The suspension point of the pendulum is free to rotate in any direction. The effect of coriolis force on the motion of the pendulum is to induce a rotation of the plane of oscillation. We will consider the motion of the pendulum at an latitude λ. The coordinate axis are chosen such that the z-axis is along the local verticle. The origin is chosen at the point where the pendulum is at rest in equilibrium and x-y axis are chosen in the horizontal plane. We assume the amplitude of oscillations to be small and let T be the tension in the wire which has a length l. For small oscillations T can be resolved into components as,

$$T = \left(-T\frac{x}{l}, -T\frac{y}{l}, T\right)$$

The tension provides the restoring force proportional to the distance along the x and y directions. The rotational velocity ω at the latitude λ has components given in equation $\omega = (0, \omega \sin\lambda, \omega \cos\lambda)$.

The coriolis force produces small velocity components in the x and y directions and there is no motion in the z-direction, thus,

$$v = (\dot{x}, \dot{y}, 0)$$

The equation of motion of the Foucault pendulum $a_{rot} = \left(\frac{dv}{dt}\right)_{rot}$ can be written as,

$$m\ddot{r} = mg + T - 2m(\omega \times v)$$

as,

$$\ddot{x} \simeq -\frac{T}{m}\frac{x}{l} + 2\dot{y}\omega \sin\lambda$$
$$\ddot{x} \simeq -\omega_o^2 x + 2\omega \sin\lambda\, \dot{y}$$
$$\ddot{y} \simeq \omega_o^2 y + 2\omega \sin\lambda\, \dot{x}$$

Where,

$$\omega_o = \sqrt{\frac{T}{ml}}$$

These equations can be solved by multiplying the second equation by i and adding the two,

$$(\dot{x} + i\dot{y}) + \omega_o^2 (x + iy) + 2i\omega \sin\lambda(\dot{x} + i\dot{y}) = 0$$

The solution can be assumed to be of the form,

$$(\dot{x} + i\dot{y}) = \eta e^{-i\alpha t}$$

Substituting in equation ($(\dot{x} + i\dot{y}) + \omega_o^2 (x + iy) + 2i\omega \sin\lambda(\dot{x} + i\dot{y}) = 0$), the general solution can be written as,

$$\eta(t) = e^{-i\omega \sin\lambda t}\left\{Ae^{i\beta t} + Be^{-i\beta t}\right\}$$

where, $\beta = \sqrt{\omega_o^2 + \omega^2 \sin^2\lambda}$

If the earth is not rotating $\omega = 0$, $\beta = \omega_o$ i.e. the pendulum oscillates with the frequency $\omega_o = \sqrt{\frac{r}{ml}}$.

$$\eta(t) \simeq e^{-i\omega \sin\lambda t}\left\{Ae^{i\omega_o t} + Be^{-i\omega_o t}\right\}$$

By equating the real and imaginary parts, we can easily obtain the solutions for x and y.

Motion of a Rigid Body

A rigid body is defined as a collection of particles which can be taken to have a discrete or continuation distribution such that the inter-particle distances within the rigid body remain fixed and do not vary. It is obvious that the rigid body so defined is an idealized concept just like the definition of a point particle. It is a useful concept and will be used in our treatment of the motion of a rigid body. The transition from the distribution of discrete particles to the volume distribution of continuous distribution of mass is simply achieved by replacing the summation over discrete particles to volume integration over pdV where p is the mass density and dv is the volume element.

To describe the motion of a rigid body we will use two coordinate systems: one an inertial frame $\Sigma_{inertial}$ and the second a frame fixed in the body called the $\Sigma_{body\ fixed}$. A rigid body has six degrees of freedom. These are taken to be the coordinates of the centre of mass (3 in number) which is also taken to coincide with the origin of the body-fixed axis. The orientation of the rigid body is fixed by choosing three suitable independent angles which fix the orientation of the body-fixed coordinates. These angles are normally chosen to be what are known as the Eulerian angles. Thus we have six coordinates to fix the orientation and the centre of mass of a rigid body.

Coordinate Systems of a Moving Rigid Body

Let the origin of the body-fixed coordinate system coincides with the centre of mass of the rigid body. Its position with respect to the inertial coordinate system be given by the vector R. Let P be any point in the rigid body whose position vector with respect to the body-fixed axis be given by with components (x_1, x_2, x_3). The position vector of P with respect to the inertial frame is given by the vector r_{in}.

We now consider an infinitesimal displacement of the rigid body. It consists of two parts: an infinitesimal translation of the centre of mass represented by the infinitesimal vector dr. The second part consists of an infinitesimal rotation about the centre of mass resulting in an infinitesimal change in the position vector of P which arises due to an infinitesimal rotation by an angle dφ and is given by dr=dφ x r. We thus get,

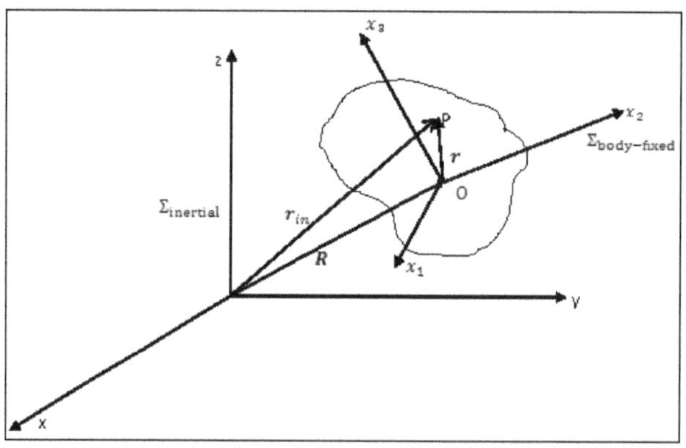

$$dr_{in} = dR + d\phi \times r$$

The displacement occurs in an infinitesimal time interval dt. Thus:

$$v = V + \omega \times r$$

where v is the velocity of the centre of mass of the body and is called the 'translation' velocity of the rigid body ω is the angular velocity of the rotation of the body and its direction is along the axis of rotation. v is the velocity of the point P in the body with respect to the inertial axis. Thus the velocity of any point in the rigid body can be expressed as a sum of translation velocity and the velocity of rotation.

We will now prove an important result: 'The angular velocity of rotation of the body-fixed coordinate system is independent of the choice of the body-fixed coordinate system'. To see this, let us shift the origin of the body-fixed axis from O to o' and the vector OO' by the vector a. Let v' be the new velocity of the point O' and ω' be the new angular velocity of rotation along the new axis of rotation. We would then have,

$$v = v' + \omega' \times r'$$

where r' is the position vector of the point P with respect to the new body-fixed coordinate system. From equation ($v = v + \omega \times r$) we have,

$$v = v + \omega \times (r' + a)$$

comparing ($v = v' + \omega' \times r'$) and above equation we get,

$$v' = v + \omega \times a$$

$$\omega' = \omega$$

Thus if the velocity v of the C.M. of the rigid body at any given instant is perpendicular to the rotation velocity ω for some choice of the origin, then v' and ω' are also perpendicular. In that case from ($v = v + \omega \times r$) the velocity of all points in the body are perpendicular to the angular velocity ω. We can then choose suitable origin such that can be made equal to and thereby from above equation is zero. Thus the motion of a rigid body at that instant is 'pure rotation' about an axis through this new choice of the origin. Such axis is called the 'instantaneous axis of rotation'.

It is however, more convenient to take the origin of the body-fixed frame to be at the centre of mass so that in general, during the motion of a rigid body, both the magnitude and the direction of the angular velocity ω vary.

The Inertia Tensor

Consider the rigid body to be compound of n particles of mass m_α, α = 1,2,...n with the position vector given by r_α with respect to the body-fixed axis. The velocity v_α of the mass m_α with respect to the inertial frame is given by,

$$v_\alpha = v + \omega \times r_\alpha$$

The translation velocity v and the rotational velocity ω are the same for every point in the body. The kinetic energy of the αth particle is given by,

$$T_\alpha = \frac{1}{2} m_\alpha v_\alpha^2$$

$$= \frac{1}{2} m_\alpha (v + \omega + r_\alpha)$$

$$= \frac{1}{2} m_\alpha v^2 + m_\alpha v \cdot \omega \times r_\alpha + \frac{1}{2} m_\alpha (\omega + r_\alpha)^2$$

The total kinetic energy of the rigid body is obtained by summing over all the particles.

$$T = \frac{1}{2}\sum_{\alpha} m_\alpha v^2 \sum_{\alpha} m_\alpha v^2 v \cdot \omega \times r_\alpha + \frac{1}{2}\sum_{\alpha} m_\alpha (\omega + r_\alpha)^2$$

$$= \frac{1}{2}\sum_{\alpha} m_\alpha v^2 + v \cdot \omega \times \sum_{\alpha} m_\alpha r_\alpha + \frac{1}{2}\sum_{\alpha} m_\alpha (\omega + r_\alpha)^2$$

$$= \frac{1}{2} M v^2 + v.\omega \times M r_\alpha + \frac{1}{2}\sum_{\alpha} m_\alpha (\omega + r_\alpha)^2$$

Since the origin of the body fixed axis is taken at the centre of mass of the rigid body,

$$\sum_{\alpha} m_\alpha r_\alpha = MR = 0$$

The kinetic energy of the rigid body can now be written as the sum of two terms, the translational energy and rotational kinetic energy. Thus,

$$T = T_{trans} + T_{rot}$$

where,

$$T_{trans} = \frac{1}{2} M v^2$$

and,

$$T_{rot} = \frac{1}{2}\sum_{\alpha} m_\alpha (\omega + r_\alpha)^2$$

using the identity,

$$(A \times B)^2 = A^2 B^2 - (A \times B)^2$$

and expressing the vector r_α in terms of its components $x_{\alpha 1}$, $x_{\alpha 2}$, $x_{\alpha 3}$,in the body-fixed system, the rotational kinetic energy can be written as,

$$T_{rot} = \frac{1}{2}\sum m_\alpha \left\{ \left(\sum_i \omega_i^2\right)\left(\sum_j x_{\alpha j}^2\right) - \left(\sum_i \omega_i x_{ai}\right)\left(\sum_j \omega_j x_{aj}\right) \right\}$$

Writing ω_i^2 as $\sum_j \omega_i \omega_j \delta_{ij}$ we can write above equation as,

$$T_{rot} = \frac{1}{2}\sum_{i.j} \omega_i \omega_j \sum_{\alpha} m_\alpha \left[\delta_{ij} \sum_k x_{\alpha k}^2 - x_{\alpha i} x_{\alpha j} \right]$$

which can be written in the form,

$$T_{rot} = \frac{1}{2}\sum_{i,j} I_{ij} \omega_i \omega_j$$

where,

$$I_{i,j}=\sum_{\alpha} m_{\alpha}\left[\delta_{i,j}\sum_{k} x_{\alpha k}^{2}-x_{\alpha i}x_{\alpha j}\right]$$

$I_{i,j}$ is called the 'Inertia Tenor' of the body. Clearly $I_{i,j}$ is symmetric in the i and j indices. $I_{i,j}$ is a tensor of rank 2 as can be seen from its definition above, notice $\delta_{I,j}$ is a symmetric tensor of rank two, $\sum_k x_{\alpha k}^2$ is a scalar and $x_{\alpha i}x_{\alpha j}$ transforms like a product of two vectors. The symmetric tensor $I_{i,j}$ has six independent components, we can write this in the form of a 3 x 3 matrix as,

$$I=\begin{bmatrix} I_{11} & I_{12} & I_{13} \\ I_{21} & I_{22} & I_{23} \\ I_{31} & I_{23} & I_{33} \end{bmatrix}$$

The components $I_{i,j}$ can be evaluated using equation ($I_{i,j}=\sum_{\alpha} m_{\alpha}\left[\delta_{i,j}\sum_{k} x_{\alpha k}^{2}-x_{\alpha i}x_{\alpha j}\right]$). For example:

$$I_{11}=\sum m_{\alpha}\left[x_{\alpha 1}^{2}+x_{\alpha 2}^{2}+x_{\alpha 3}^{2}\right]=\sum m_{\alpha}\left(x_{\alpha 2}^{2}+x_{\alpha 3}^{2}\right)$$

Likewise,

$$I_{22}=\sum m_{\alpha}\left[x_{\alpha 1}^{2}+x_{\alpha 3}^{2}\right];\ I_{33}=\sum m_{\alpha}\left(x_{\alpha 1}^{2}+x_{\alpha 2}^{2}\right)$$
$$I_{12}=\sum m_{\alpha}x_{\alpha 1}x_{\alpha 2}$$

and so on. Thus,

$$I=\sum_{\alpha} m_{\alpha}\begin{bmatrix} \left(x_{\alpha_2}^{2}+x_{\alpha 3}^{2}\right) & -x_{\alpha 1}x_{\alpha 2} & -x_{\alpha 1}x_{\alpha 3} \\ -x_{\alpha 1}x_{\alpha 2} & \left(x_{\alpha 1}^{2}+x_{\alpha 3-}^{2}\right) & -x_{\alpha 2}x_{\alpha 3} \\ -x_{\alpha 1}x_{\alpha 3} & -x_{\alpha 2}x_{\alpha 3} & \left(x_{\alpha 1}^{2}+x_{\alpha 2}^{2}\right) \end{bmatrix}$$

The diagonal elements I_{11}, I_{22}, I_{33}, are called the moment of inertia about the x_1, x_2, x_3, and axis respectively. If we have a continuous distribution of mass in the rigid body, the sum in ($I_{i,j}=\sum_{\alpha} m_{\alpha}\left[\delta_{i,j}\sum_{k} x_{\alpha k}^{2}-x_{\alpha i}x_{\alpha j}\right]$) is replaced by an integral namely,

$$I_{i,j}=\int_{V} p(r)\left[\delta_{ij}\sum_{k} x_{k}^{2}-x_{i}x_{j}\right]dV$$

Where p(r) is the volume mass density dV and is the volume element $dV = dx_1\, dx_2\, dx_3$.

The inertia tensor $I_{i,j}$ being a symmetric tensor of rank-2 can be reduced to a diagonal form by a suitable choice of the axis x_1, x_2, x_3. Corresponding to these Principal Axis of Inertia the diagonal components of the tensor are called Principal Moments of Inertia denoted by I_1, I_2 and I_3. The kinetic energy of rotation of a rigid body about the Principal axis of rotation can be written as,

$$T_{rot}=\frac{1}{2}\left(I_{1}\omega_{1}^{2}+I_{2}\omega_{2}^{2}+I_{3}\omega_{3}^{2}\right)$$

If the rigid body has some symmetry, then its centre of mass and principal axis also has the same symmetry. When there are two axis of symmetry, i.e. $I_1 = I_2 \neq I_3$ body is called a 'symmetric top'. When the body has spherical symmetry i.e. $I_1 = I_2 = I_3$ the rigid body is called a 'spherical top'. An 'asymmetric top' has no axis of symmetry and I_1, I_2 and I_3 are all different.

Angular Momentum of a Rigid Body

Angular momentum of a system depends on the point with respect to which it is defined. For a rigid body, the most appropriate choice of this point is the origin of the rotating body-fixed axis. The origin of the body-fixed axis is in turn is taken to be the centre of mass of the body.

For a rigid body, the angular momentum (L) is the product of the moment of inertia and the angular velocity: $L = I\omega$. For a point of mass, angular momentum can be expressed as the product of linear momentum and the radius (r): $L = mvr$. L is measured in units of kilograms-meters^2 per second or more commonly joule-seconds. The law of conservation of angular momentum can be stated that the angular momentum of a system of objects is conserved if there is not external net torque acting on the system.

Analogous to Newton's law ($F = \Delta(mv)/\Delta t$) there is a rotational counterpart for rotational motion: $t = \Delta L/\Delta t$, or torque is the rate of change of angular momentum.

Consider the example of a child who runs tangential to the edge of a playground merry-go-round with a velocity v_0 and jumps on while the merry-go-round is at rest. The only external forces are that of gravity and the contact forces provided by the support bearings, neither of which causes a torque because they are not applied to cause a horizontal rotation. Treat the child's mass as a point of mass and the merry-go-round as a disc with a radius R and mass M. From the conservation law, the total angular momentum of the child before the interaction is equal to the total angular momentum of the child and merry-go-round after the collision: $mrv_0 = mrv' + I\omega$, where r is the radial distance from the center of the merry-go-round to the place where the child hits. If the child jumps on the edge, ($r = R$) and the angular velocity for the child after the collision can be substituted for the linear velocity, $mRv_0 = mR(R\omega) + (1/2)MR^2$. If the values for the masses and the initial velocity of the child are given, the final velocity of the child and merry-go-round can be calculated.

A single object may have a change in angular velocity due to the conservation of angular momentum if the distribution of the mass of the rigid body is altered. For example, when a figure skater pulls in her extended arms, her moment of inertia will decrease, causing an increase in angular velocity. According to the conservation of angular momentum, $I_0(\omega_0) = I_f(\omega_f)$ where I_0 is the moment of inertia of the skater with arms extended, I_f is her moment of inertia with her arms close to her body, ω_0 is her original angular velocity, and ω_f is her final angular velocity.

References

- Hamilton's-Principle-and-Lagrange's-Equation: phy.duke.edu, Retrieved 23, June 2020
- Coriolis-force, science: britannica.com, Retrieved 30, March 2020
- Rotational-motion-of-a-rigid-body, classical-mechanics, physics: cliffsnotes.com, Retrieved 10, August 2020

3

Principles and Concepts of Quantum Mechanics

Quantum mechanics is a fundamental theory of physics that describes physical properties of nature at atomic and subatomic scale. The various principles and equations studied in relation to quantum mechanics are Heisenberg's uncertainty principle, generalized uncertainty principle, Heisenberg equations of motion, etc. These topics are crucial for a complete understanding of the subject.

Quantum mechanics is the science dealing with the Behaviour of matter and light on the atomic and subatomic scale. It attempts to describe and account for the properties of molecules and atoms and their constituents—electrons, protons, neutrons, and other more esoteric particles such as quarks and gluons. These properties include the interactions of the particles with one another and with electromagnetic radiation (i.e., light, X-rays, and gamma rays).

The behaviour of matter and radiation on the atomic scale often seems peculiar, and the consequences of quantum theory are accordingly difficult to understand and to believe. Its concepts frequently conflict with common-sense notions derived from observations of the everyday world. There is no reason, however, why the behaviour of the atomic world should conform to that of the familiar, large-scale world. It is important to realize that quantum mechanics is a branch of physics and that the business of physics is to describe and account for the way the world—on both the large and the small scale—actually is and not how one imagines it or would like it to be.

The study of quantum mechanics is rewarding for several reasons. First, it illustrates the essential methodology of physics. Second, it has been enormously successful in giving correct results in practically every situation to which it has been applied. There is, however, an intriguing paradox. In spite of the overwhelming practical success of quantum mechanics, the foundations of the subject contain unresolved problems—in particular, problems concerning the nature of measurement. An essential feature of quantum mechanics is that it is generally impossible, even in principle, to measure a system without disturbing it; the detailed nature of this disturbance and the exact point at which it occurs are obscure and controversial. Thus, quantum mechanics attracted some of the ablest scientists of the 20th century, and they erected what is perhaps the finest intellectual edifice of the period.

Photoelectric Effect

The photoelectric effect is a phenomenon in which electrons are ejected from the surface of a metal when light is incident on it. These ejected electrons are called photoelectrons. It is important to note that the emission of photoelectrons and the kinetic energy of the ejected photoelectrons is dependent on the frequency of the light that is incident on the metal's surface. The process through which photoelectrons are ejected from the surface of the metal due to the action of light is commonly referred to as photoemission.

The photoelectric effect occurs because the electrons at the surface of the metal tend to absorb energy from the incident light and use it to overcome the attractive forces that bind them to the metallic nuclei. An illustration detailing the emission of photoelectrons as a result of the photoelectric effect is provided below.

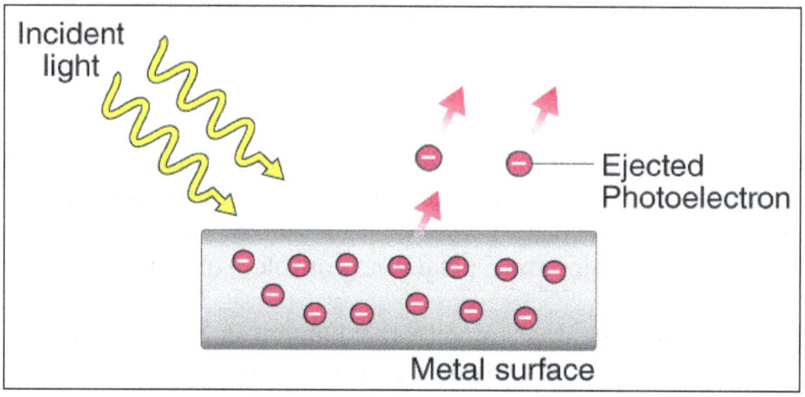

The Concept of Photons

The photoelectric effect cannot be explained by considering light as a wave. However, this phenomenon can be explained by the particle nature of light, in which light can be visualized as a stream of particles of electromagnetic energy. These 'particles' of light are called photons. The energy held by a photon is related to the frequency of the light via Planck's equation:

$$E = h\nu = hc/\lambda$$

Where,

- E denotes the energy of the photon.
- h is Planck's constant.
- ν denotes the frequency of the light.
- c is the speed of light (in a vacuum).
- λ is the wavelength of the light.

Thus, it can be understood that different frequencies of light carry photons of varying energies. For example, the frequency of blue light is greater than that of red light (the wavelength of blue light is much shorter than the wavelength of red light). Therefore, the energy held by a photon of blue light will be greater than the energy held by a photon of red light.

Threshold Energy for the Photoelectric Effect

For the photoelectric effect to occur, the photons that are incident on the surface of the metal must carry sufficient energy to overcome the attractive forces that bind the electrons to the nuclei of the metals. The minimum amount of energy required to remove an electron from the metal is called the threshold energy (denoted by the symbol Φ). For a photon to possess energy equal to the threshold energy, its frequency must be equal to the threshold frequency (which is the minimum frequency of light required for the photoelectric effect to occur). The threshold frequency is usually denoted by the symbol ν_{th} and the associated wavelength (called the threshold wavelength) is denoted by the symbol λ_{th}. The relationship between the threshold energy and the threshold frequency can be expressed as follows.

$$\Phi = h\nu_{th} = hc/\lambda_{th}$$

Relationship between the Frequency of the Incident Photon and the Kinetic Energy of the Emitted Photoelectron

Therefore, the relationship between the energy of the photon and the kinetic energy of the emitted photoelectron can be written as follows:

$$E_{photon} = \Phi + E_{electron}$$

$$\Rightarrow h\nu = h\nu_{th} + \frac{1}{2}m_e v^2$$

Where,

- E_{photon} denotes the energy of the incident photon, which is equal to $h\nu$.
- Φ denotes the threshold energy of the metal surface, which is equal to $h\nu_{th}$.
- $E_{electron}$ denotes the kinetic energy of the photoelectron, which is equal to ½mev2 (me = mass of electron = $9.1*10^{-31}$ kg).

If the energy of the photon is less than the threshold energy, there will be no emission of photoelectrons (since the attractive forces between the nuclei and the electrons cannot be overcome). Thus, the photoelectric effect will not occur if $\nu < \nu_{th}$. If the frequency of the photon is exactly equal to the threshold frequency ($\nu = \nu_{th}$), there will be an emission of photoelectrons, but their kinetic energy will be equal to zero. An illustration detailing the effect of the frequency of the incident light on the kinetic energy of the photoelectron is provided below.

From the image, it can be observed that:

- The photoelectric effect does not occur when the red light strikes the metallic surface because the frequency of red light is lower than the threshold frequency of the metal.
- The photoelectric effect occurs when green light strikes the metallic surface and photoelectrons are emitted.
- The photoelectric effect also occurs when blue light strikes the metallic surface. However, the kinetic energies of the emitted photoelectrons are much higher for blue light than for green light. This is because blue light has a greater frequency than green light.

It is important to note that the threshold energy varies from metal to metal. This is because the attractive forces that bind the electrons to the metal are different for different metals. It can also be noted that the photoelectric effect can also take place in non-metals, but the threshold frequencies of non-metallic substances are usually very high.

Einstien's Contributions Towards the Photoelectric Effect

The photoelectric effect is the process that involves the ejection or release of electrons from the surface of materials (generally a metal) when light falls on them. The photoelectric effect is an important concept that enables us to clearly understand the quantum nature of light and electrons.

After continuous research in this field, the explanation for the photoelectric effect was successfully explained by Albert Einstein. He concluded that this effect occurred as a result of light energy being carried in discrete quantized packets. For this excellent work, he was honoured with the Nobel prize in 1921.

According to Einstein, each photon of energy E is,

$$E = h\nu$$

Where,

- E = Energy of photon in joule.
- h = planks constant (6.626×10^{-34} J.s).
- ν = frequency of photon in Hz.

Properties of the Photon

- For a photon, all the quantum numbers are zero.
- A photon does not have any mass, charge and they are not reflected in a magnetic and electric field.
- The photon moves in the speed of light at empty space.
- During the interaction of matters with radiation, radiation behaves as it is made up of small particles called photons.
- Photons are virtual particles. The photon energy is directly proportional to its frequency and inversely proportional to its wavelength.
- The momentum and energy of the photons are related as given below:

  $$E = p.c$$

Where,

- p = magnitude of the momentum.

- c = speed of light.

The phenomenon of metals releasing electrons when they are exposed to the light of the appropriate frequency is called the photoelectric effect, and the electrons emitted during the process are called photoelectrons.

The law of conservation of energy forms the basis for the photoelectric effect.

Minimum Condition for Photoelectric Effect

Threshold Frequency (γ_{th})

It is the minimum frequency of the incident light or radiation that will produce a photoelectric effect i.e. ejection of photoelectrons from a metal surface is known as threshold frequency for the metal. It is constant for a specific metal but may be different for different metals.

If γ = frequency of incident photon and γ_{th}= threshold frequency, then,

- If $\gamma < \gamma_{Th}$, there will be no ejection of photoelectron and, therefore, no photoelectric effect.
- If $\gamma = \gamma_{Th}$, photoelectrons are just ejected from the metal surface, in this case, the kinetic energy of the electron is zero.
- If $\gamma > \gamma_{Th}$, then photoelectrons will come out of the surface along with kinetic energy.

Threshold Wavelength (λ_{th})

During the emission of electrons, a metal surface corresponding to the greatest wavelength to incident light is known threshold wavelength.

$$\lambda_{th} = c/\gamma_{th}$$

For wavelengths above this threshold, there will be no photoelectron emission. For λ = wavelength of the incident photon, then:

- If $\lambda < \lambda Th$, then the photoelectric effect will take place and ejected electron will possess kinetic energy.
- If $\lambda = \lambda Th$, then just photoelectric effect will take place and kinetic energy of ejected photoelectron will be zero.
- If $\lambda > \lambda Th$, there will be no photoelectric effect.

Work Function or Threshold Energy (Φ)

The minimal energy of thermodynamic work that is needed to remove an electron from a conductor to a point in the vacuum immediately outside the surface of the conductor is known as work function/threshold energy:

$$\Phi = h\gamma_{th} = hc/\lambda_{th}$$

The work function is the characteristic of a given metal. If E = energy of an incident photon, then:

- If $E < \Phi$, no photoelectric effect will take place.
- If $E = \Phi$, just photoelectric effect will take place but the kinetic energy of ejected photoelectron will be zero.
- If E > photoelectron will be zero.
- If $E > \Phi$, the photoelectric effect will take place along with possession of the kinetic energy by the ejected electron.

Photoelectric Effect Formula

According to the Einstein explanation of the photoelectric effect is: The energy of photon = energy needed to remove an electron + kinetic energy of the emitted electron is,

$$h\nu = W + E$$

Where,

- h is Planck's constant.
- ν is the frequency of the incident photon.
- W is a work function.
- E is the maximum kinetic energy of ejected electrons: $1/2\ mv^2$.

Laws Governing the Photoelectric Effect

For a light of any given frequency; $(\gamma > \gamma_{Th})$ photoelectric current is directly proportional to the intensity of light. For any given material, there is a certain minimum (energy) frequency, called threshold frequency, below which the emission of photoelectrons stops completely, no matter how high is the intensity of incident light.

The maximum kinetic energy of the photoelectrons is found to increase with the increase in the frequency of incident light, provided the frequency $(\gamma > \gamma_{Th})$ exceeds the threshold limit. The maximum kinetic energy is independent of the intensity of light. The photo-emission is an instantaneous process.

Experimental Study of Photoelectric Effect

The given set up D experiment is used to study the photoelectric effect experimentally. In an evacuated glass tube. Two zinc plates C and D are enclosed. Plates C acts as anode and D acts as a photosensitive plate.

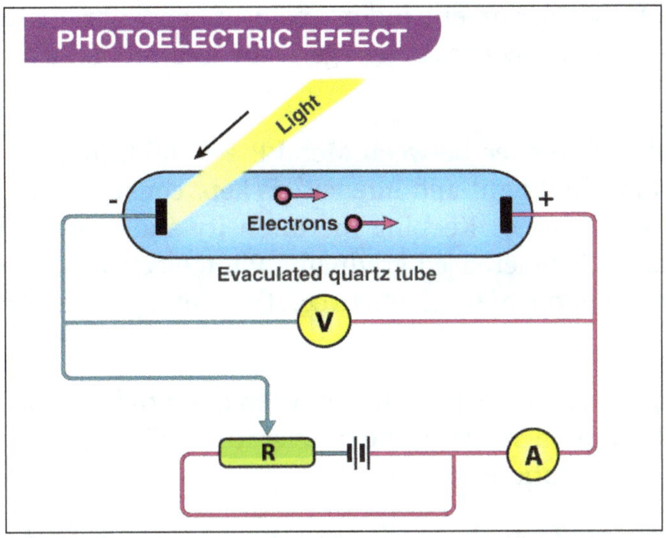

Two plates are connected to a battery B and ammeter A. If the radiation is incident on the plate D through a quartz window W electrons are ejected out of the plate and current flows in the circuit this is known as photocurrent. Plate C can be maintained at desired potential (+ve or – ve) with respect to plate D.

Characteristics of Photoelectric Effect

- The threshold frequency varies with material, it is different for different materials.
- The photoelectric current is directly proportional to the light intensity.
- The kinetic energy of the photoelectrons is directly proportional to the light frequency.
- The stopping potential is directly proportional to the frequency and the process is instantaneous.

Factors Affecting Photoelectric Effect

With the help of this apparatus, we will now study the dependence of the photoelectric effect on the following factors:

- The intensity of incident radiation.
- A potential difference between metal plate and collector.
- Frequency of incident radiation.
- Effects of Intensity of Incident Radiation on Photoelectric EffectL: The potential difference between the metal plate and collector and frequency of incident light is kept constant and the intensity of light is varied: The electrode C i.e. collecting electrode is made positive with respect to D (metal plate). For a fixed value of frequency and the potential between the metal plate and collector, the photoelectric current is noted in accordance with the intensity of incident radiation.

It shows that photoelectric current and intensity of incident radiation both are proportional to each other the photoelectric current gives an account of the number of photoelectrons ejected per sec.

- Effects of Potential Difference between Metal Plate and Collector on Photoelectric Effect: The frequency of incident light and intensity is kept constant and the potential difference between the plates is varied: Keeping the intensity and frequency of light constant, the positive potential of C is increased gradually. Photoelectric current increases when there is a positive increase in the potential between the metal plate and collector up to a characteristic value.

  There is no change in photoelectric current when potential increased higher than the characteristic value for any increase in the accelerating voltage. This maximum value of the current is called as saturation current.

- Effect of Frequency on Photoelectric Effect: The intensity of light is kept constant and the frequency of light is varied: For a fixed intensity of incident light, variation in the frequency of incident light produces linear in the variation cut off potential/stopping potential of the metal. It shown cut off potential (Vc) is linearly proportional to the frequency of incident light.

  The kinetic energy of the photoelectrons increases directly proportionally to the frequency of incident light to completely stop the photoelectrons. We should reverse and increase the potential between the metal plate and collector in (negative value) so the emitted photoelectron can't reach the collector.

Einstein's Photoelectric Equation

According to Einstein's theory on photoelectric effect is, when a photon collides inelastically with electrons, the photon is absorbed completely or partially by the electrons. So if an electron in a metal absorbs a photon of energy, it uses the energy in the following ways.

Some energy Φ_0 is used to making the surface electron free from the metal. It is known as the work function of the material. Rest energy will appear as kinetic energy (K) of the emitted photoelectrons. Einstein's Photoelectric Equation explains the following concepts:

- The frequency of the incident light is directly proportional to the kinetic energy of the electrons and the wavelengths of incident light are inversely proportional to the kinetic energy of the electrons.
- If $\gamma = \gamma_{th}$ or $\lambda = \lambda_{th}$ then $v_{max} = 0$.
- $\gamma < \gamma_{th}$ or $\lambda > \lambda_{th}$: There will be no emission of photoelectrons.
- The intensity of the radiation or incident light refers to the number of photons in the light beam. More intensity means more photons and vice-versa. Intensity has nothing to do with the energy of the photon. Therefore, the intensity of the radiation is increased, the rate of emission increases but there will be no change in kinetic energy of electrons. With an increasing number of emitted electrons, the value of photoelectric current increases.

Applications of Photoelectric Effect

- Used to generate electricity in Solar Panels. These panels contain metal combinations that allow electricity generation from a wide range of wavelengths.
- Motion and Position Sensors: In this case, a photoelectric material is placed in front of a UV or IR LED. When an object is placed in between the Light-emitting diode (LED) and sensor, light is cut off and the electronic circuit registers a change in potential difference.
- Lighting sensors such as the ones used in smartphones enable automatic adjustment of screen brightness according to the lighting. This is because the amount of current generated via the photoelectric effect is dependent on the intensity of light hitting the sensor.
- Digital cameras can detect and record light because they have photoelectric sensors that respond to different colors of light.
- X-Ray Photoelectron Spectroscopy (XPS): This technique uses x-rays to irradiate a surface and measure the kinetic energies of the emitted electrons. Important aspects of the chemistry of a surface can be obtained such as elemental composition, chemical composition, the empirical formula of compounds and chemical state.
- Photoelectric cells are used in burglar alarms.

The Compton Effect

The Compton effect (also called Compton scattering) is the result of a high-energy photon colliding with a target, which releases loosely bound electrons from the outer shell of the atom or molecule. The scattered radiation experiences a wavelength shift that cannot be explained in terms of classical wave theory, thus lending support to Einstein's photon theory. The effect was first demonstrated in 1923 by Arthur Holly Compton (for which he received a 1927 Nobel Prize). The effect is important because it demonstrates that light cannot be explained purely as a wave phenomenon.

Thomson scattering, the classical theory of an electromagnetic wave scattered by charged particles, cannot explain low intensity shift in wavelength. Classically, light of sufficient intensity for the electric field to accelerate a charged particle to a relativistic speed will cause radiation-pressure recoil and an associated Doppler shift of the scattered light, but the effect would become arbitrarily small at sufficiently low light intensities regardless of wavelength. Light must behave as if it consists of particles in order to explain the low-intensity Compton scattering. Compton's experiment convinced physicists that light can behave as a stream of particles whose energy is proportional to the frequency.

The Experiment

A schematic diagram of the apparatus used by Compton is shown in the Figure below. A graphite target was bombarded with monochromatic x-rays and the wavelength of the scattered radiation was measured with a rotating crystal spectrometer. The intensity was determined by a movable ionization chamber that generated a current proportional to the x-ray intensity. Compton measured the dependence of scattered x-ray intensity on wavelength at three different scattering angles of 45°, 90°, and 135°.

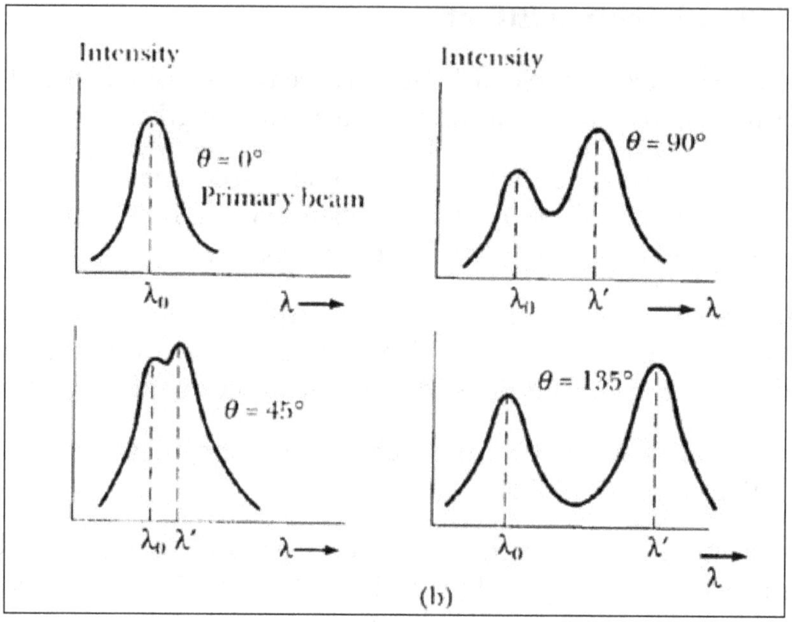

(b)

The experimental intensity vs. wavelength plots observed by Compton for the above three scattering angles show two peaks, one at the wavelength λ of the incident x-rays and the other at a longer wavelength λ_c'.

Physics

Compton effect is the decrease in energy (increase in wavelength) of an X-ray or gamma ray photon, when it interacts with matter. Because of the change in photon energy, it is an inelastic scattering process. Inverse Compton scattering also exists, where the photon gains energy (decreasing in wavelength) upon interaction with matter. Compton used a combination of three fundamental formulas representing the various aspects of classical and modern physics, combining them to describe the quantum behavior of light:

- Light as a particle, as noted in the photoelectric effect.
- Relativistic dynamics: special theory of relativity.
- Trigonometry: law of cosines.

We can think of the photon as losing an amount of energy in the collision that is the same as the kinetic energy (KE) gained by the electron, although actually separate photons are involved. If the initial photon has the frequency ν associated with it, the scattered photon has the lower frequency ν', where,

Loss in photon energy = gain in electron energy:

$$h\nu - h\nu' = KE$$

and the momentum of a massless particle is related to its energy by the formula,

$$E = pc$$

since the energy of a photon is \(h\nu\), its momentum is,

$$p = \frac{E}{c} = \frac{h\nu}{c}$$

Momentum, unlike energy is a vector quantity that incorporates direction as well as magnitude, and in the collision momentum must be conserved in each of two mutually perpendicular directions (when more than two bodies participate in a collision, momentum must be conserved in each of three mutually perpendicular directions). The directions we choose here are that of the original photon and one perpendicular to it in the plane containing the electron and the scattered photon.

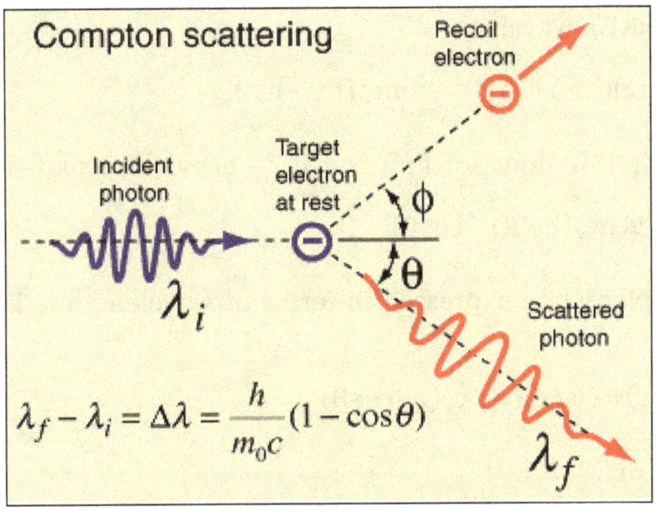

The initial photon momentum is hν/c, the scattered photon momentum is hν′/c, and the initial and final electron momenta are respectively o and p. In the original photon direction.

Initial momentum = final momentum

$$h\nu / c + o = (h\nu' / c)\cos\theta + p\cos\phi$$

and perpendicular to this direction,

Initial momentum= final momentum

$$o = (h\nu'/c) \sin \theta - p\sin \phi$$

From Equation ($h\nu - h\nu' = KE$), ($h\nu / c + o = (h\nu' / c)\cos\theta + p\cos\phi$), ($o = (h\nu'/c)\sin \theta - p \sin\phi$), we can find a formula that relates the wavelength difference between initial and scattered photons with the angle θ between their directions, both of which are readily measurable quantities (unlike the energy and momentum of the recoil electron).

The first step is to multiply both the Equations by c and rewrite them as,

$$(pc)\cos\phi = h\nu - (h\nu')\cos\theta$$

$$(pc)\sin\phi = (h\nu')\sin\theta$$

By squaring each of these equations and adding the new ones together, the angle ϕ is eliminated, leaving,

$$p^2c^2 = (h\nu)^2 - 2(h\nu)(h\nu')\cos\theta + (h\nu')^2$$

Next we equate the two expressions for the total energy of a particle,

$$E = KE + mc^2$$

$$E = \sqrt{m^2c^4 + p^2c^2}$$

From equation ($h\nu - h\nu' = KE$), we have,

$$p^2c^2 = (h\nu)^2 - 2(h\nu)(h\nu') + (h\nu')^2 + 2mc^2(h\nu - h\nu')$$

Substituting this value of p^2c^2 in Equation ($p^2c^2 = (h\nu)^2 - 2(h\nu)(h\nu')\cos\theta + (h\nu')^2$), we finally obtain,

$$2mc^2(h\nu - h\nu') = 2(h\nu)(h\nu')(1 - \cos\theta)$$

This relationship is simpler when expressed in terms of wavelength λ. Dividing Equation above by $2h^2c^2$,

$$mc / h(\nu / c - \nu' / c) = (\nu / c)(\nu' / c)(1 - \cos\theta)$$

and so, $\nu/c = 1/\lambda$ and $\nu'/c = 1/\lambda'$

$$mc / h(1/\lambda - 1/\lambda') = (1 - \cos\theta) / \lambda\lambda'$$
$$\lambda - \lambda' = h / mc(1 - \cos\theta)$$

Equation above gives the change in wavelength expected for a photon that is scattered through the angle θ by a particle mass mm. This change is independent of the wavelength λ of incident photon. The quantity,

$$\lambda_c = \frac{h}{mc}$$

is called the Compton wavelength of the scattering particle. For an electron $\lambda_c = 2.426 \times 10^{-12}$m. In terms of λ_c, equation ($\lambda - \lambda' = h / mc(1 - \cos\theta)$) becomes,

$$\lambda - \lambda' = \lambda c(1 - \cos\theta)$$

The Compton wavelength gives the scale of the wavelength change of the incident photon. From Equation above we note that the greatest wavelength change possible corresponds to $\theta = 180°$, when the wavelength change will be twice the Compton wavelength λ_c. Because $\lambda_c = 0.00243$nm for an electron, and even less for other particles owing to their larger rest masses, the maximum wavelength change in the Compton effect is 0.004852 nm.

If the photon is of lower energy, but still has sufficient energy (in general a few eV, right around the energy of visible light), it can eject an electron from its host atom entirely (a process known as the photoelectric effect), instead of undergoing Compton scattering. Higher energy photons (~MeV) may be able to bombard the nucleus and cause an electron and a positron to be formed, a process called pair production.

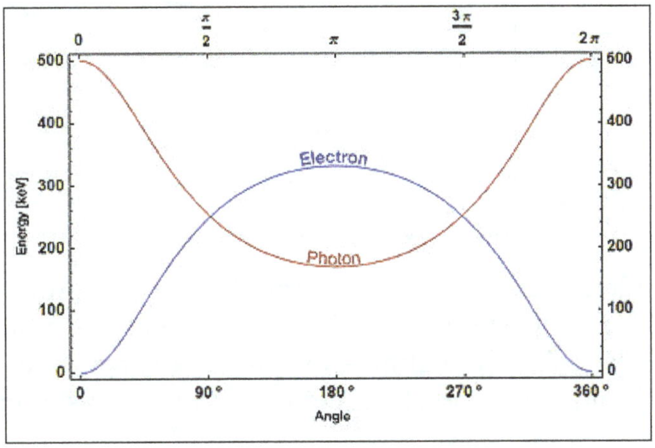

Applications

Compton scattering is of prime importance to radiobiology, as it happens to be the most probable interaction of high energy X rays with atomic nuclei in living beings and is applied in radiation therapy. In material physics, Compton scattering can be used to probe the wave function of the electrons in matter in the momentum representation. Compton scattering is an important effect in gamma spectroscopy which gives rise to the Compton edge, as it is possible for the gamma rays to scatter out of the detectors used. Compton suppression is used to detect stray scatter gamma rays to counteract this effect.

Young's Double-Slit Experiment

Young's double-slit experiment uses two coherent sources of light placed at a small distance apart, usually, only a few orders of magnitude greater than the wavelength of light is used. Young's double-slit experiment helped in understanding the wave theory of light which is explained with the help of a diagram. A screen or photo-detector is placed at a large distance 'D' away from the slits as shown.

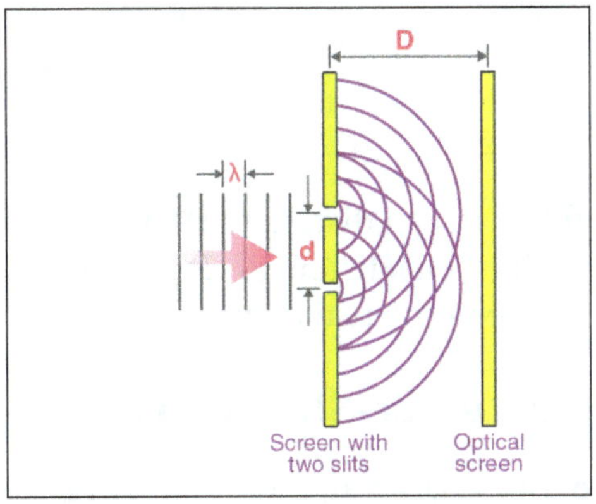

The original Young's double-slit experiment used diffracted light from a single source passed into two more slits to be used as coherent sources. Lasers are commonly used as coherent sources in the modern-day experiments.

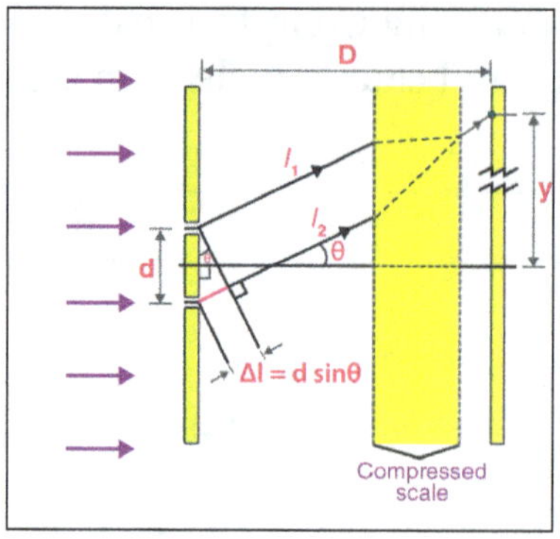

Each source can be considered as a source of coherent light waves. At any point on the screen at a distance 'y' from the centre, the waves travel distances l_1 and l_2 to create a path difference of Δl at the point in question. The point approximately subtends an angle of θ at the sources (since the distance D is large there is only a very small difference of the angles subtended at sources).

Derivation of Young's Double-Slit Experiment

Consider a monochromatic light source's' kept at a considerable distance from two slits s_1 and s_2. S is equidistant from s_1 and s_2. s_1 and s_2 behave as two coherent sources, as both bring derived from S. The light passes through these slits and falls on a screen which is at a distance 'D' from the position of slits s_1 and s_2. 'd' be the separation between two slits.

If s_1 is open and s_2 is closed, the screen opposite to s_1 is closed, only the screen opposite to s_2 is illuminated. The interference patterns appear only when both slits s_1 and s_2 are open.

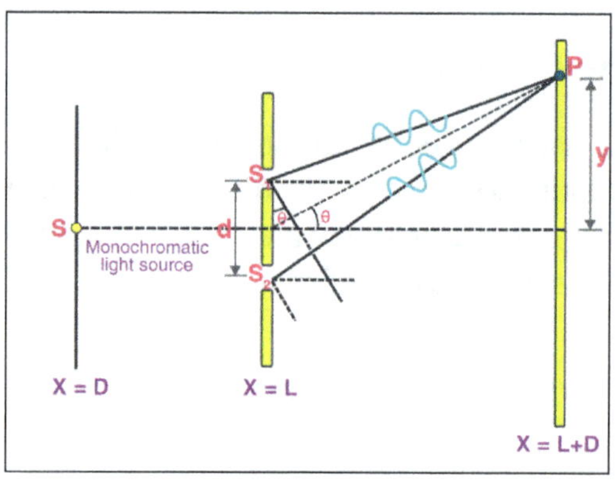

When the slit separation (d) and the screen distance (D) are kept unchanged, to reach P the light waves from s_1 and s_2 must travel different distances. It implies that there is a path difference in Young's double slit experiment between the two light waves from s_1 and s_2.

Approximations in Young's Double-Slit Experiment

- Approximation 1:D > > d: Since D > > d, the two light rays are assumed to be parallel, then the path difference.
- Approximation 2:d/λ >> 1: Often, d is a fraction of a millimetre and λ is a fraction of a micrometre for visible light.

Under these conditions θ is small, thus we can use the approximation sin θ approx tan θ = γ/D.

∴ path difference, Δz = γ/D

This is the path difference between two waves meeting at a point on the screen. Due to this path difference in Young's double slit experiment, some points on the screen are bright and some points are dark.

Position of Fringes in Young's Double-Slit Experiment

Position of Bright Fringes

For maximum intensity at P,

$$\Delta z = n\lambda \ (n = 0, \pm 1, \pm 2, \ldots.)$$

$$\text{Or } d \sin\theta = n\lambda \ (n = 0, \pm 1, \pm 2, \ldots.)$$

The bright fringe for n = 0 is known as the central fringe. Higher order fringes are situated symmetrically about the central fringe. The position of nth bright fringe is given by,

$$y\ (\text{bright}) = (n\lambda \backslash d)D \ (n = 0, \pm 1, \pm 2, \ldots.)$$

Position of Dark Fringes

For minimum intensity at P,

$$\Delta z = (2n-1)\frac{n\lambda}{2}(n = \pm 1, \pm 2, \ldots..)$$

$$d \sin\theta = (2n-1)\frac{n\lambda}{2}$$

The first minima are adjacent to the central maximum on either side. We will obtain the position of dark fringe as,

$$y\,\text{dark} = \frac{(2n-1)\lambda D}{2d}(n = \pm 1, \pm 2, \ldots..)$$

Fringe Width

Distance between two adjacent bright (or dark) fringes is called the fringe width.

$$\beta z\frac{n\lambda D}{d} - \frac{(n-1)\lambda D}{d} = \frac{\lambda D}{d}$$

$$\Rightarrow \beta \propto \lambda$$

If the apparatus of Young's double slit experiment is immersed in a liquid of refractive index (u), then wavelength of light and hence fringe width decreases 'u' times.

$$\beta^1 = \frac{\beta}{\mu}$$

If a white light is used in place of monochromatic light, then coloured fringes are obtained on the screen with red fringes larger in size that of violet.

Angular Width of Fringes

Let the angular position of nth bright fringe is θ_n and because of its small value $\tan\theta_n \approx \theta_n$,

$$\tan\theta_n = \frac{\gamma n}{D} \approx \theta_n = \frac{\gamma n}{D}$$

$$\Rightarrow \theta_n = \frac{n\gamma D/d}{D} = \frac{n\gamma}{d}$$

Similarly, the angular position of (n+1)[th] bright fringe is, $\theta n+1$, then,

$$\theta_{n+1} = \frac{\gamma n+1}{D} = \frac{(n+1)\lambda D/d}{D} = \frac{(n+1)\lambda}{d}$$

$\therefore$ Angular width of a fringe in Young's double slit experiment is given by,

$$\theta = \theta_{n+1} - \theta_n = \frac{(n+1)\lambda}{d} - \frac{n\lambda}{d} = \frac{\lambda}{d}\theta = \frac{\lambda}{d}$$

We know that,

$$\beta = \frac{\lambda D}{d}$$

$$\Rightarrow \theta = \frac{\lambda}{d} = \frac{\beta}{D}$$

Angular width is independent of 'n' i.e angular width of all fringes are same.

Maximum Order of Interference Fringes

The position of nth order maxima on the screen is,

$$\gamma = \frac{n\lambda D}{d}; n = 0, \pm 1, \pm 2, \ldots$$

But 'n' values cannot take infinitely large values as it would violate 2nd approximation. i.e θ is small (or)

$$y << D$$

$$\Rightarrow \frac{\gamma}{D} = \frac{n\lambda}{d} << 1$$

Hence, the above formula for interference maxima is applicable when $n << \frac{d}{\lambda}$. When 'n' value becomes comparable to $\frac{d}{\lambda}$, path difference can no longer be given by $\frac{d\gamma}{D}$, Hence for maxima, path difference = $n\lambda$,

$$\Rightarrow d\sin\theta = n\lambda$$

$$\Rightarrow n = \frac{d\sin\theta}{\lambda} n_{max} = \left[\frac{d}{\lambda}\right]$$

The above represents box function or greatest integer function. Similarly, the highest order of interference minima,

$$n_{min} = \left[\frac{d}{\lambda} + \frac{1}{2}\right]$$

Shape of Interference Fringes in YDSE

From the given YDSE diagram, the path difference from the two slits is given by,

$$s_2p - s_1p \approx d\sin\theta(\text{constant})$$

The above equation represents a hyperbola with its two foci as s_1 and s_2.

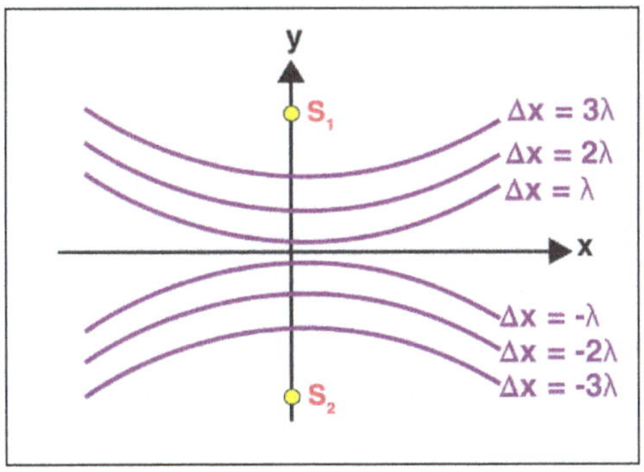

The interference pattern we get on the screen is a section of a hyperbola when we revolve hyperbola about the axis s_1s_2. If the fringe will represent 1st minima, the fringe will represent 1st maxima, it represents central maxima. If the screen is yz plane, fringes are hyperbolic with a straight central section.

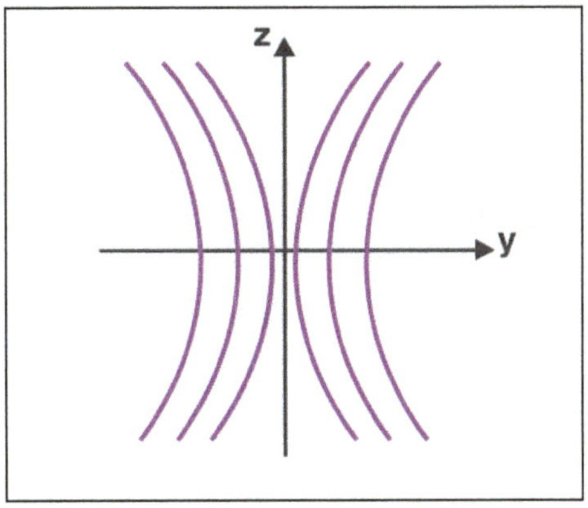

If the screen is xy plane, the fringes are hyperbolic with a straight central section.

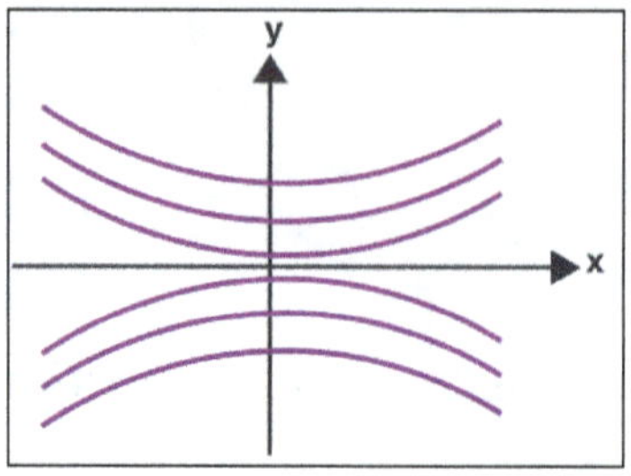

Intensity of Fringes in Young's Double-Slit Experiment

For two coherent sources s_1 and s_2, the resultant intensity at point p is given by,

$$I = I1 + I2 + 2\sqrt{(I1 . I2)} \cos \varphi$$

Putting, I1 = I2 = Io (Since, d<<<D)

$$I = Io + Io + 2\sqrt{(Io.Io)} \cos \varphi$$

$$I = 2Io + 2 (Io) \cos \varphi$$

I = 2Io (H cos s φ)

$$I = 4I_0 \cos^2\left(\frac{\phi}{2}\right)$$

For Maximum Intensity

$$\cos\frac{\phi}{2} = \pm 1 \frac{\phi}{2} = n\pi, n = 0, \pm 1, \pm 2, ...$$

$$\text{or } \varphi = 2n\pi$$

Phase difference: $\varphi = 2n\pi$

then, path difference: $\Delta x = \frac{\lambda}{2\pi}(2n\pi) = n\lambda$. The intensity of bright points are maximum and given by Imax = 4Io.

For Minimum Intensity

$$\cos\frac{\phi}{2} = 0 \frac{\phi}{2}$$

$$= \left(n - \frac{1}{2}\right)\pi \text{ where} \left(n = \pm 1, \pm 2, \pm 3, ...\right)$$

$$\varphi = (2n - 1)\pi$$

Phase Difference: $\varphi = (2n - 1)\pi$,

$$\frac{2\pi}{\lambda}\Delta x = (2n - 1)$$

$$\pi \Delta x = (2n - 1)\frac{\lambda}{2}$$

Thus, intensity of minima is given by Imin = 0. If,

$$I_1 \neq I_2, I_{min} \neq 0.$$

Special Cases

Rays not Parallel to Principal Axis

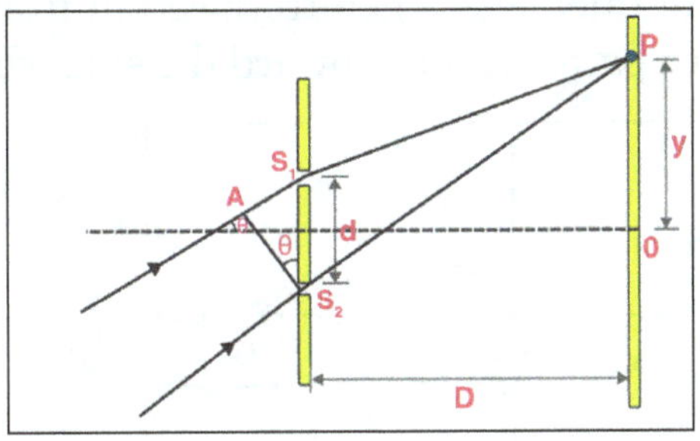

From the above diagram, Path difference:

$$\Delta x = (AS_1 + S_1P) - S_2P$$
$$\Delta x = AS_1 - (S_2P - S_1P)\Delta x$$
$$= d\sin\theta - \frac{4d}{D}$$

- For maxima: $\Delta x = n\lambda$
- For minima: $\Delta x = (2n-1)\frac{\lambda}{2}$

Using this we can calculate different positions of maxima and minima.

Source Placed Beyond the Central Line

If the source is placed a little above or below this centre line, the wave interaction with S_1 and S_2 has a path difference at a point P on the screen,

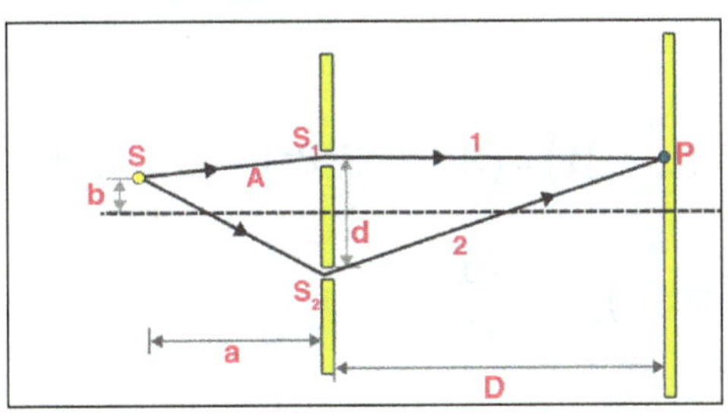

Δx = (distance of ray 2) – (distance of ray 1)

$$= (S\,S_2 + S_2P) - (S\,S_1 + S_1P)$$
$$= (S\,S_2 + S\,S_1) + (S_2P - S_1P)$$

$$= bd/a + yd/D \rightarrow (*)$$

We know,

- $\Delta x = n\lambda$ for maximum.
- $\Delta x = (2n - 1)\lambda/2$ for minimum.

Displacement of Fringes in YDSE

When a thin transparent plate of thickness't' is introduced in front of one of the slits in Young's double slit experiment, the fringe pattern shifts toward the side where the plate is present.

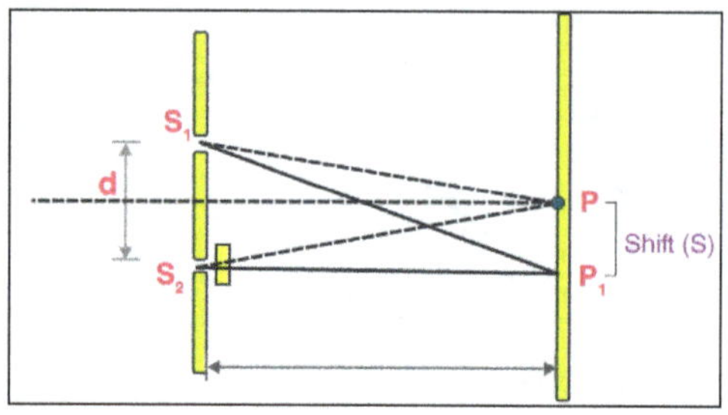

The dotted lines denote the path of the light before introducing the transparent plate. The solid lines denote the path of the light after introducing a transparent plate.

- Path difference before introducing the plate $\Delta x = S_1P - S_2P$.
- Path difference after introducing the plate $\Delta x_{new} = S_1P^1 - S_2P^1$.

The path length,

$$S_2P^1 = (S_2P^1 - t)_{air} + t_{plate}$$

$$= (S_2P^1 - t)_{air} + (\mu t)_{plate}$$

Where 'μt' is Optical Path,

$$S_2P^1{}_{air} + (\mu - 1)t$$

Then we get,

$$(\Delta x)_{new} = S_1P^1_{air} - (S_2P^1_{air} + (\mu - 1)t) = (S_1P^1 - S_2P^1)_{air} - (\mu - 1)t$$

$$(\Delta x)_{new} = d\sin\theta - (\mu - 1)t(\Delta x)_{new} = \frac{\gamma d}{D} - (\mu - 1)t$$

$$(\Delta x)_{new} = \frac{\gamma d}{D} - (\mu - 1)t$$

Then,

$$y = \underset{(1)}{\underset{\downarrow}{\frac{\Delta x D}{d}}} + \underset{(2)}{\underset{\downarrow}{\frac{D}{d}}}[(\mu - 1)t]$$

Term (1) defines the position of a bright or dark fringe, the term (2) defines the shift occurred in the particular fringe due to the introduction of a transparent plate.

Constructive and Destructive Interference

For constructive interference, the path difference must be an integral multiple of the wavelength. Thus for a bright fringe to be at 'y',

$n\lambda = y\, dD$

Or, ynth = $n\lambda\, Dd$

Where n = ±0,1,2,3.....

The 0th fringe represents the central bright fringe. Similarly, the expression for a dark fringe in Young's double slit experiment can be found by setting the path difference as: $\Delta l = (n+12)\lambda$. This simplifies to, $yn = (n+12)\lambda Dd$.

Note that these expressions require that θ be very small. Hence yD needs to be very small. This implies D should be very large and y should be small. This, in turn, requires that the formula works best for fringes close to the central maxima. In general, for best results, dD must be kept as small as possible for a good interference pattern.

The Young's double slit experiment was a watershed moment in scientific history because it firmly established that light indeed behaved as a wave. The Double Slit Experiment was later conducted using electrons, and to everyone's surprise, the pattern generated was similar as expected with light. This would forever change our understanding of matter and particles, forcing us to accept that matter like light also behaves like a wave.

Heisenberg's Uncertainty Principle

One of the important contents of the theory of quantum mechanics is the uncertainty principle, developed by W. Heisenberg in 1927. Stated in simple terms, it says that if the x-coordinate of the position of a quantum mechanical particle is known to an accuracy, Δx, then the x-component of the corresponding momentum variable cannot be determined to an accuracy better than $\Delta p_x \approx h / \Delta x$. In other words, one cannot determine precisely and simultaneously the values of both the members of particular pairs of physical variables, known as canonically conjugate to each other, which define a given atomic system. The x- (or y- or z-) component of the position coordinate of a particle and the corresponding component of momentum, p_x (or p_y, p_z), a component L_z of angular momentum and its angular coordinate φ (in the xy-plane), the energy E of a particle and the time t at which the energy is measured – are some of the examples of canonically conjugate pairs. Thus, expressed in quantitative terms, the uncertainty principle states that,

$$\Delta x \Delta p_x \geq \hbar, \quad \Delta L_z \Delta \phi \geq \hbar, \quad \Delta E\, \Delta t \geq \hbar$$

The fact that h is so small makes the uncertainty principle relevant primarily to the systems of atomic size.

Generalized Uncertainty Principle

The generalized uncertainty principle indicates that measuring one observable may provide you some information as well on its incompatible partners, which in fact has no conflict with the well-known interpretation of complementarity.

The statement on Heisenberg's uncertainty principle in terms of the uncertainties, Δx and Δp in the measurement of position and momentum variables, the component of angular momentum and its canonically conjugate variable. Here we shall study how this principle is derived formally in general terms.by introducing the uncertainties ΔA and ΔB, associated with the measurement of the dynamical variables A and B in a given state. We shall use the generalized uncertainty relation to find out the uncertainty relation between the components of orbital angular momentum we shall also derive the expression for the wave packet which represents the minimum uncertainty in position and momentum variables.

Generalized Uncertainty Principle

We made the statement on Heisenberg's uncertainty principle in terms of the uncertainties, Δx and Δp in the measurement of position and momentum variables. Having studied the postulates of quantum mechanics, in particular, the commutation relation of quantum mechanical operators, we are now in a position to derive the generalized uncertainty principle general statement of the uncertainty principle for any pair of non- commuting Hermitian operators, $\hat{A}$ and $\hat{B}$ use this in specific examples.

Root Mean Square Deviation

As one of the postulates of quantum mechanics, we have learnt that an expectation value of an operator is an average over several measurements, each of which will deviate from the average value. The average of these deviations would of course be zero. But the average of the square of these deviations - mean square deviations would be non-zero. The square root of the mean-square deviation, called root mean square-deviation can be taken as measure of spread in the values measured. These spreads ΔA and ΔB in the measured values of the observed quantities A and B are called the uncertainties.

Derivation of the General Statement

Thus the uncertainties ΔA and ΔB associated with the measurement of A and B in a given quantum mechanical state, ψ, are defined as,

$$(\Delta A)^2 = \left\langle \left(\hat{A} - <\hat{A}>\right)^2 \right\rangle = \left\{ <\hat{A}^2> - 2<\hat{A}><\hat{A}> + <\hat{A}>^2 \right\}$$

$$= <\hat{A}^2> - <\hat{A}>^2$$

The uncertainty depends on the state ψ in which the system is. Similarly, for the other Hermitian operator $\hat{B}$,

$$(\Delta B)^2 = \left\langle \left(\hat{B} - <\hat{B}>\right)^2 \right\rangle = <\hat{B}^2> - <\hat{B}>^2$$

Let us define,

$$\hat{E} = \hat{A} - <\hat{A}> \text{ and } \hat{F} = \hat{B} - <\hat{B}>$$

and consider,

$$|\phi\rangle = \left(\hat{E} + i\lambda\hat{F}\right)|\Psi\rangle$$

where, for simplicity, λ is real. Since are $\hat{A}$ and $\hat{B}$ are Hermitian so are $\hat{E}$ and $\hat{F}$. Now,

$$\langle\phi|\phi\rangle = \left\langle\psi|\left(\hat{E} - i\lambda\hat{F}\right)\left(\hat{E} + i\lambda\hat{F}\right)|\psi\right\rangle \geq 0$$

$$\text{or, } \left\langle\psi|\hat{E}^2 + \lambda^2\hat{F}^2 + i\lambda\left[\hat{E} + \hat{F}\right]|\psi\right\rangle \geq 0$$

where we note,

$$\left[\hat{E},\hat{F}\right] = \left[\left(\hat{A} - <\hat{A}>\right),\left(\hat{B} - <\hat{B}>\right)\right] = \left[\hat{A},\hat{B}\right]$$

Rewriting Equation ($\langle\phi|\phi\rangle = \left\langle\psi|\left(\hat{E} - i\lambda\hat{F}\right)\left(\hat{E} + i\lambda\hat{F}\right)|\psi\right\rangle \geq 0$ or $\left\langle\psi|\hat{E}^2 + \lambda^2\hat{F}^2 + i\lambda\left[\hat{E} + \hat{F}\right]|\psi\right\rangle \geq 0$) in terms of the operators, $\hat{A}$ and $\hat{B}$, we have,

$$\left\langle\psi\left|\left(\hat{A} - <\hat{A}>\right)^2\right|\psi\right\rangle + \lambda^2\left\langle\psi\left|\left(\hat{B} - <\hat{B}>\right)^2\right|\psi\right\rangle + i\lambda\left\langle\psi\left|\left[\hat{A},\hat{B}\right]\right|\psi\right\rangle \geq 0$$

The left hand side of above equation has its minimum value when its derivative with respect to λ is zero. This gives,

$$2\lambda(\Delta B)^2 + i\left\langle\psi\left|\left[\hat{A},\hat{B}\right]\right|\psi\right\rangle = 0$$

$$\text{or } \lambda = -\frac{\left\langle\psi\left|\left[\hat{A},\hat{B}\right]\right|\psi\right\rangle}{2(\Delta B)}$$

Substituting this value of λ from both the above equations, we have,

$$\left\langle\psi\left|\left(\hat{A} - <\hat{A}>^2\right)\right|\psi\right\rangle - \frac{\left|\left\langle\psi\left|\left[\hat{A},\hat{B}\right]\right|\psi\right\rangle^2\right|}{4(\Delta B)^2} + \frac{\left|\left\langle\psi\left|\left[\hat{A},\hat{B}\right]\right|\psi\right\rangle^2\right|}{2(\Delta B)^2} \geq 0$$

Abve equation simplifies to,

$$(\Delta A)^2(\Delta B)^2 \geq \frac{1}{4}\left|\left\langle\psi\left|\left[\hat{A},\hat{B}\right]\right|\psi\right\rangle\right|^2$$

$$\text{or } \geq \frac{1}{4}\left|\frac{1}{i}\left\langle\psi\left|\left[\hat{A},\hat{B}\right]\right|\psi\right\rangle\right|^2 \geq \frac{1}{4}\left|\left\langle\psi\left|\hat{C}\right|\psi\right\rangle\right|^2$$

In above equation, we introduce the relation:

$$\left[\hat{A},\hat{B}\right] = i\hat{C}$$

where $\hat{C}$ is Hermitian. We get finally,

$$(\Delta A)(\Delta B) \geq \frac{1}{2} < \hat{C} >$$

Above equation gives the general statement of the uncertainty principle for any pair of non- commuting Hermitian operators, $\hat{A}$ and $\hat{B}$, which satisfy the relation Equation ($\left[\hat{A}, \hat{B}\right] = i\hat{C}$).

Uncertainty Relation between pair of Canonically Conjugate Operators

In particular, for a pair of conjugate operators $\hat{A}$ and $\hat{B}$, which satisfy the commutation relation,

$$\left[\hat{A}, \hat{B}\right] = i\hbar$$

we have, according to Equation ($(\Delta A)(\Delta B) \geq \frac{1}{2} < \hat{C} >$),

$$(\Delta A)(\Delta B) \geq \frac{1}{2}\hbar$$

And we know that a position coordinate variable x and its canonically conjugate momentum variable p satisfy the relation, Equation ($\left[\hat{A}, \hat{B}\right] = i\hbar$). Accordingly,

$$\Delta x \Delta p \geq \hbar / 2$$

This is the precise statement of the uncertainty principle in terms of the measurement of position and momentum observables, x and p_x. The generalization to three dimensions is straight forward, i.e.,

$$(\Delta x)(\Delta p_x) \geq \frac{1}{2}\hbar, \quad \text{and} \quad (\Delta y)(\Delta p_y) \geq \frac{1}{2}\hbar, \quad \text{and} \quad (\Delta z)(\Delta p_z) \geq \frac{1}{2}\hbar$$

Corresponding to the commutation relation,

$$\left[\hat{x}_i, \hat{p}_j\right] = i\hbar\delta_{ij} \quad i, j = 1, 2, 3$$

whereas we have,

$$\left[\hat{x}_i, \hat{x}_j\right] = 0 \quad \text{and} \quad \left[\hat{p}_i, \hat{p}_j\right] = 0 \quad i.j = 1, 2, 3$$

Uncertainty Relation between the components of Angular Momentum Operators

We shall encounter in our future course of study several other examples of non-commuting Hermitian operators. Consider, for example, the orbital angular momentum of a particle, defined as; $\vec{L} = \vec{r} \times \vec{p}$ about a fixed point. Expressed in Cartesian components,

$$L_x = yp_z - zp_y,$$

etc. in cyclic order.

Replacing the position and momentum variables as operators in quantum mechanics, we can write the commutation relations between its components, as for example,

$$\left[\hat{L}_x,\hat{L}_y\right]=\left[(\hat{y}\hat{p}_z-\hat{z}\hat{p}_z),(\hat{z}\hat{p}_x-\hat{x}\hat{p}_z)\right]$$
$$=\left[\hat{y}\hat{p}_z-\hat{z}\hat{p}_x\right]-\left[\hat{y}\hat{p}_z-\hat{x}p_z\right]-\left[\hat{z}\hat{p}_y,\hat{z}\hat{p}_x\right]+\left[\hat{z}\hat{p}_y,\hat{x}\hat{p}_z\right]$$

Note that in the two middle terms, none of the operators are accompanied by its canonically conjugate partners. As a result, these terms would vanish. The first term can be expressed as $\hat{y}\hat{p}_x\left[\hat{p}_z,\hat{z}\right]=-i\hbar\hat{y}\hat{p}_x$ Similarly, the last term gets simplified as $\left[\hat{z},\hat{p}_z\right]\hat{x}\hat{p}_y=i\hbar\hat{x}\hat{p}_y$. Putting these two terms together, we have,

$$\left[\hat{L}_x,\hat{L}_y\right]=i\hbar\left(\hat{x}\hat{p}_y-\hat{y}\hat{p}_x\right)=i\hbar\hat{L}_z$$

The commutators of other two components can be worked out in the same way. Thus we have,

$$\left[\hat{L}_x,\hat{L}_y\right]=i\hbar\hat{L}_z,\quad \left[\hat{L}_y,\hat{L}_z\right]=i\hbar\hat{L}_x,\quad \left[\hat{L}_z,\hat{L}_x\right]=i\hbar\hat{L}_y$$

Applying the generalized uncertainty relation, Equation ($(\Delta A)(\Delta B)\geq\frac{1}{2}\hbar$) for the components of angular momentum operators we get,

$$(\Delta L_x)(\Delta L_y)\geq\frac{\hbar}{2}|<L_z>|$$

i.e., the uncertainty in the product of measurement of the x- and y-components of orbital angular is determined by the expectation value of the z-component of the orbital momentum. Similar relations hold for other pairs of the angular momentum components as well.

Linear Vector Spaces in Quantum Mechanics

We have observed that most operators in quantum mechanics are linear operators. This is fortunate because it allows us to represent quantum mechanical operators as matrices and wave functions as vectors in some linear vector space. Since computers are particularly good at performing operations common in linear algebra (multiplication of a matrix times a vector, etc.), this is quite advantageous from a practical standpoint.

In an n-dimensional space we may expand any vector Ψ as a linear combination of basis vectors,

$$\Psi=\sum_{i=1}^{n}a_i\Psi_i$$

For a general vector space, the coefficients a_i may be complex; thus one should not be too quick to draw parallels to the expansion of vectors in three-dimensional Euclidean space. The coefficients a_i are referred to as the "components" of the state vector Ψ , and for a given basis, the components of a vector specify it completely. The components of the sum of two vectors are the sums of the components. If $\Psi_a = \sum a_i \Psi_i$ and $\Psi_b = \sum b_i \Psi_i$ then,

$$\Psi_a + \Psi_b = \sum_i (a_i + b_i)\Psi_i$$

and similarly,

$$\lambda\Psi_a = \sum_i (\lambda a_i)\Psi_i$$

The scalar product of two vectors is a complex number denoted by,

$$(\Psi_b, \Psi_a) = (\Psi_a, \Psi_b)^*$$

where we have used the standard linear-algebra notation. If we also require that,

$$(\Psi_a, \lambda\Psi_b) = \lambda(\Psi_a, \Psi_b)$$

then it follows that,

$$(\lambda\Psi_a, \Psi_b) = \lambda^*(\Psi_a, \Psi_b)$$

We also require that,

$$(\Psi_a + \Psi_b + \Psi_c) = (\Psi_a, \Psi_b) + (\Psi_a, \Psi_c)$$

If the scalar product vanishes (and if neither vector in the product is the null vector) then the two vectors are orthogonal.

Generally the basis is chosen to be orthonormal, such that,

$$(\hat{\Psi}_i, \hat{\Psi}_j) = \delta_{ij}$$

In this case, we can write the scalar product of two arbitrary vectors as,

$$\begin{aligned}(\Psi_a, \Psi_b) &= \left(\sum_i a_i \hat{\Psi}_i, \sum_i b_j \hat{\Psi}_j\right) \\ &= \sum_i \sum_j a_i^* b_j (\hat{\Psi}_i, \hat{\Psi}_j) \\ &= \sum_i a_i^* b_i\end{aligned}$$

This can also be written in vector notation as,

$$(\Psi_a, \Psi_b) = (a_1^* a_2^* \ldots a_n^*)\begin{pmatrix} b_1 \\ b_2 \\ \vdots \\ b_n \end{pmatrix}$$

It is useful at this point to introduce Dirac's bra-ket notation. We define a ``bra" as,

$$\langle \Psi_a | = \left(a_1^* a_2^* \ldots a_n^*\right)$$

and a "ket" as,

$$|\Psi_a\rangle = \begin{pmatrix} a_1 \\ a_2 \\ \vdots \\ a_n \end{pmatrix}$$

A bra to the left of a ket implies a scalar product, so,

$$\langle \Psi_a | \Psi_b \rangle = (\Psi_a | \Psi_b)$$

Sometimes in superficial treatments of Dirac notation, the symbol $\langle \Psi_a | \Psi_b \rangle$ is defined alternatively as,

$$\langle \Psi_a | \Psi_b \rangle = \int \Psi_a^*(x) \Psi_b(x) dx$$

This is equivalent to the above definition if we make the connections $a_i = \Psi_a(x)$ and $b_i = \Psi_b(x)$. This means that our basis vectors are every possible value of $\underline{x}$. Since is continuous, the sum is replaced by an integral. Often only the subscript of the vector is used to denote a bra or ket; we may have written the above equation as,

$$\langle a | b \rangle = \int \Psi_a^*(x) \Psi_b(x) dx$$

An operator $\hat{A}$ can be characterized by its effect on the basis vectors. The action of $\hat{A}$ on a basis vector $\hat{\Psi}_j$ yields some new vector $\underline{\hat{\Psi}'_j}$ which can be expanded in terms of the basis vectors so long as we have a complete basis set.

$$\hat{A}\hat{\Psi}_j = \hat{\Psi}'_j = \sum_i^n \hat{\Psi}_i A_{ij}$$

If we know the effect of $\hat{A}$ on the basis vectors, then we know the effect of $\hat{A}$ on any arbitrary vector because of the linearity of $\hat{A}$.

$$\begin{aligned} \Psi_b = \hat{A}\Psi_a = \hat{A}\sum_j a_j \hat{\Psi}_j = \sum_j a_j \hat{A}\hat{\Psi}_j = \sum_j \sum_i a_j \hat{\Psi}_i A_{ij} \\ = \sum_i \hat{\Psi}_i \left(\sum_j A_{ij} a_j \right) \end{aligned}$$

Or,

$$b_i = \sum_j A_{ij} a_j$$

This may be written in matrix notation as,

$$\begin{pmatrix} b_1 \\ b_2 \\ \vdots \\ b_n \end{pmatrix} = \begin{pmatrix} A_{11} & A_{12} & \cdots & A_{1n} \\ A_{21} & A_{22} & \cdots & A_{2n} \\ \vdots & \vdots & & \vdots \\ A_{n1} & A_{n2} & \cdots & A_{nn} \end{pmatrix} \begin{pmatrix} a_1 \\ a_2 \\ \vdots \\ a_n \end{pmatrix}$$

We can obtain the coefficients A_{ij} by taking the inner product of both sides of equation $\hat{A}\hat{\Psi}_j = \hat{\Psi}_j' = \sum_i^n \hat{\Psi}_i A_{ij}$ with $\hat{\Psi}_i$, yielding,

$$\begin{aligned}\left(\hat{\Psi}_i \hat{A}\hat{\Psi}_j\right) &= \left(\hat{\Psi}_i, \sum_k^n \hat{\Psi}_k A_{kj}\right) \\ &= \sum_k^n A_{kj}\left(\hat{\Psi}_i, \hat{\Psi}_k\right) \\ &= A_{ij}\end{aligned}$$

since $\left(\hat{\Psi}_i, \hat{\Psi}_k\right) = \delta_{ik}$ due to the orthonormality of the basis. In bra-ket notation, we may write,

$$A_{ij} = \left\langle i \middle| \hat{A} \middle| j \right\rangle$$

where i and j denote two basis vectors. This use of bra-ket notation is consistent with its earlier use if we realize that $\hat{A}|j\rangle$ is just another vector $|j'\rangle$.

It is easy to show that for a linear operator $\hat{A}$, the inner product $\left(\Psi_a, \hat{A}\Psi_b\right)$ for two general vectors (not necessarily basis vectors) Ψ_a and Ψ_b is given by,

$$\left(\Psi_a, \hat{A}\Psi_b\right) = \sum_i \sum_j a_i^* A_{ij} b_j$$

or in matrix notation,

$$\left(\Psi_a, \hat{A}\Psi_b\right) = \left(a_1^* a_2^* \ldots a_n^*\right) \begin{pmatrix} A_{11} & A_{12} & \cdots & A_{1n} \\ A_{21} & A_{22} & \cdots & A_{2n} \\ \vdots & \vdots & & \vdots \\ A_{n1} & A_{n2} & \cdots & A_{nn} \end{pmatrix} \begin{pmatrix} b_1 \\ b_2 \\ \vdots \\ b_n \end{pmatrix}$$

By analogy to equation $\langle \Psi_a | \Psi_b \rangle = \int \Psi_a^*(x) \Psi_b(x) dx$ we may generally write this inner product in the form,

$$\left(\Psi_a, \hat{A}\Psi_b\right) = \left\langle a \middle| \hat{A} \middle| b \right\rangle = \int \Psi_a^*(x) \hat{A} \Psi_b(x) dx$$

Previously, we noted that $(\Psi_a, \Psi_b) = (\Psi_b, \Psi_a)^*$, or $\langle a | b \rangle = \langle b | a \rangle^*$. Thus we can see also that,

$$\left(\Psi_a, \hat{A}\Psi_b\right) = \left(\hat{A}\Psi_b, \Psi_a\right)^*$$

We now define the adjoint of an operator $\hat{A}$ denoted by $\hat{A}^{\dagger}$ as that linear operator for which,

$$\left(\Psi_a, \hat{A}\Psi_b\right) = \left(\hat{A}^{\dagger}\Psi_a, \Psi_b\right)$$

That is, we can make an operator act backwards into ``bra" space if we take its adjoint. With this definition, we can further see that,

$$\left(\Psi_a, \hat{A}\Psi_b\right) = \left(\hat{A}\Psi_b, \Psi_a\right)^* = \left(\Psi_b, \hat{A}^{\dagger}\Psi_a\right)^* = \left(\hat{A}^{\dagger}\Psi_a, \Psi_b\right)$$

or, in bra-ket notation,

$$\left\langle a\middle|\hat{A}\middle|b\right\rangle = \left\langle \hat{A}b\middle|a\right\rangle^* = \left\langle b\middle|\hat{A}^{\dagger}\middle|a\right\rangle^* = \left\langle \hat{A}^{\dagger}a\middle|b\right\rangle$$

If we pick $\Psi_a = \hat{\Psi}_i$ and $\Psi_b = \hat{\Psi}_j$ (i.e., if we pick two basis vectors), then we obtain,

$$\left(\hat{A}\hat{\Psi}_i, \hat{\Psi}_j\right) = \left(\hat{\Psi}_i, \hat{A}^{\dagger}\hat{\Psi}_j\right)$$

$$\left(\hat{\Psi}_j, \hat{A}\hat{\Psi}_i\right)^* = \left(\hat{\Psi}_i, \hat{A}^{\dagger}\hat{\Psi}_j\right)$$

$$A^*_{ji} = \hat{A}^{\dagger}_{ij}$$

But this is precisely the condition for the elements of a matrix and its adjoint. Thus the adjoint of the matrix representation of $\hat{A}$ is the same as the matrix representation of $\hat{A}^{\dagger}$.

This correspondence between operators and their matrix representations goes quite far, although of course the specific matrix representation depends on the choice of basis. For instance, we know from linear algebra that if a matrix and its adjoint are the same, then the matrix is called Hermitian. The same is true of the operators; if,

$$\hat{A} = \hat{A}^{\dagger}$$

then $\hat{A}$ is a Hermitian operator, and all of the special properties of Hermitian operators apply to $\hat{A}$ or its matrix representation.

Postulates of Quantum Mechanics

State Vectors and Observables of a Physical System

Postulate: Each state of a physical system is represented by a vector ket, say, $|a\rangle$, referred to as a state vector in a Hilbert space. The vector $|a\rangle$ or any scalar multiple of it corresponds to the same physical state. For convenience, it is required that $|a\rangle$ be normalized so that the vector representing the state is determined upto a constant factor.

To the above postulate is added the principle of linear superposition, according to which if $|a_1\rangle$ and $|a_2\rangle$ represent the possible state vectors then a linear combination,

$$|a\rangle = \alpha |a_1\rangle + \beta |a_2\rangle$$

is also a possible state vector, where α and β are arbitrary complex numbers. Conversely, any state may be considered as a linear superposition of two or more states. In fact, we assume that there exists an orthonormal complete set of state vectors for any state space.

You may recall that in classical mechanics, the dynamical state of a system is specified by dynamical variables such as coordinates and momenta of the particles in the system at a given instant. From this information, not only the value of any other dynamical variable such as energy can be obtained but even the state of the system at any other time can be deduced by means of the equations of motion. In quantum mechanics, however, the definition of the state of the system is significantly different from that of the classical case. Here to get information on the dynamical variables of the system, we introduce the following postulate.

Postulate: For every dynamical observable A (like energy, momentum etc.) of the system, there exists a Hermitian operator $\hat{A}$ in the Hilbert space. As stated above, any state vector can be expressed in terms of a complete set of basis vectors. These basis vectors are the eigen vectors of the Hermitian operator. The only measurable values of a physical observable are the various eigenvalues of the corresponding operator. Thus, each of the eigen values a_k (assumed, for the time being, to be discrete) of the operator has associated with it a vector, represented by $|a_k\rangle$, in the Hilbert space. Let it be an eigen vector of operator $\hat{A}$ so that,

$$\hat{A}|a_k\rangle = a_k |a_k\rangle$$

The hermitian character of the operator ensures that the eigenvalues are real. Eigen vectors corresponding to different eigen values are orthogonal, i.e.,

$$\langle a_i | a_j\rangle = 0, \quad \text{for} \quad i \neq j$$

and also normalized, i.e., $\langle a_k | a_k\rangle = 1$.

Measurement Postulates: Expectation Values and Probabilities

Postulate: If we measure the variable A in the state of the system specified by the eigen-vector $|a_k\rangle$ of the operator $\hat{A}$, we shall have the precise measurement in terms of the eigen value, a_k. If the measurement is made on large number of identical systems, each characterized by a state represented by a normalized ket $|\psi\rangle$, then the average value (expectation value) of the operator $\hat{A}$ is defined by,

$$\langle \hat{A}\rangle_{av} = \langle \psi|\hat{A}|\psi\rangle$$

In case, $|\psi\rangle$ is an eigen vector of $\hat{A}$, say, $|\psi\rangle \equiv |a\ \rangle$, where,

$$\hat{A}|a_k\rangle = \alpha_k |a_k\rangle$$

Then, $\langle \hat{A}\rangle = \alpha_k$, which is in agreement with the statement given earlier. If $|\psi\rangle$ is not an eigenvector of Â, then $|\psi\rangle$ can be expanded in terms of the eigenvectors $\{|a_k\rangle\}$ of Â, which form a complete orthonormal set of Hermitian operator. We then have,

$$|\psi\rangle = \sum_k |a_k\rangle\langle a_k|\psi\rangle$$

so that,

$$\langle\psi|\hat{A}|\psi\rangle = \sum_{k.l}\langle\psi||a_k\rangle\langle a_k|\hat{A}|a_l\rangle\langle a_l||\psi\rangle$$

$$= \sum_{k.l}\langle\psi|a_k\rangle\delta_{kl}a_k\langle a_l|\psi\rangle = \sum_k\langle\psi|a_k\rangle\langle a_k|\psi\rangle\alpha_k$$

Note that in Eq. ($\langle\psi|\hat{A}|\psi\rangle = \sum_{k.l}\langle\psi||a_k\rangle\langle a_k|\hat{A}|a_l\rangle\langle a_l||\psi\rangle$), we have twice made use of the closure relation. We thus express,

$$\langle\psi|\hat{A}|\psi\rangle = \sum_k|\langle\psi|a_k\rangle|^2\alpha_k$$

Equation above is to be interpreted as giving us the result of a large number of measurements under identical conditions or, equivalently, the result of a measurement on a large number of identical systems, each measurement yielding one or the other eigenvalue of Â. As one cannot, in general, predict which eigen-state will be obtained, we postulate that the probability of finding the state $|a_k\rangle$ is $P(|a_k\rangle) = |\langle a_k|\psi\rangle|^2$.

Thus, $\langle a_k|\psi\rangle$ may be regarded as the probability amplitude for the system to be found in $|a_k\rangle$. Indeed the concept of probability amplitude is based on the premise that $|\psi\rangle$, as expressed by Equation ($|\psi\rangle = \sum_k|a_k\rangle\langle a_k|\psi\rangle$), represents the state of the physical system.

Quantum Mechanical Operators

Postulate: Let there be a dynamical variable, as for instance a Hamiltonian, in classical mechanics, which is a function of canonically conjugate variables. The corresponding quantum mechanical operator of the system is obtained from the dynamical variable in classical mechanics by replacing the canonically conjugate variables by the corresponding quantum mechanical operators.

Consider the example of a linear (one-dimensional) harmonic oscillator in classical mechanics, where its Hamiltonian (in Cartesian co-ordinates) is given by,

$$H \equiv H(x,p) = \frac{P^2x}{2m} + \frac{1}{2}Kx^2$$

where m is the mass of the oscillator and K is a spring constant. Here the Hamiltonian is a function of the position co-ordinate, x, and the momentum, P_x, which are the canonical variables.

According to the postulate, the quantum mechanical operator corresponding to H is obtained as,

$$\hat{H} \equiv \hat{H}(\hat{x},\hat{p}_x) = \frac{\hat{p}^2 x}{2m} + \frac{1}{2}K\hat{x}^2$$

where $\hat{x}$ and $\hat{p}_x$ are Hermitian operators corresponding to x and p_x respectively.

It is important to keep in mind that quantum mechanical operators of the conjugate variables need not commute; therefore proper order of the variables has to be preserved. To make the point clear, let us consider, for example, the orbital angular momentum; $\vec{L} = \vec{r} \times \vec{p}$, where its x-component is given by,

$$L_x = yp_z - zp_y, \quad \text{cyclic,}$$

$$L_x^2 = yp_zyp_z + zp_yzp_y - yp_zzp_y - zp_yyp_z$$

So that, $= y^2p_z^2 + z^2p_y^2 - 2yp_yzp_z$.

Classically, both the expressions of above equation are equivalent. However, when the canonical variables are replaced by the operators, the corresponding expressions would not give the same result for the operator $\hat{L}_x^2$. The correct expression is obtained by replacing the variables by corresponding operators in the equation given by Equation ($L_x^2 = yp_zyp_z + zp_yzp_y - yp_zzp_y - zp_yyp_z$). Only those operators which commute can be permuted.

Another point to note is that a dynamical variable C which is the product two other dynamical variables, say, A and B, i.e., C=A B, then the Hermitian operator corresponding to C is not $\hat{A}\hat{B}$, because if these operators do not commute, i.e., $\hat{A}\hat{B} \neq \hat{B}\hat{A}$, the product $\hat{A}\hat{B}$ would then be not Hermitian. It is a simple exercise to check that a proper combination is given by $\hat{C} = \frac{1}{2}(\hat{A}\hat{B} + \hat{B}\hat{A})$, which is Hermitian.

Postulate: Any pair of canonically conjugate operators satisfies the Heisenberg commutation rules:

$$[\hat{q}_i, \hat{q}_j] = 0;$$

$$[\hat{p}_i, \hat{p}_j] = 0;$$

$$[\hat{q}_i, \hat{p}_k] = i\hbar\delta_{ik}$$

where the operator $\hat{q}_i$ represents the generalized coordinate corresponding to q_i and similarly the operator $\hat{p}_i$ is the operator corresponding the generalized momentum p_i canonically conjugate to q_i. Thus, if q_i represent the Cartesian coordinates, the p_i is the components of linear momentum. If, on the other hand, q_i represent the angle, p_i then are the components of angular momentum and so on. In this abstract formalism, the commutation relation, like,

$$[\hat{x}, \hat{p}_x] = i\hbar$$

between a position variable and its canonically conjugate momentum may be regarded as a fundamental postulate and is not to be derived . This quantum condition forms the basis of quantum mechanics.

As another example, consider the case of orbital angular momentum. Thus substitution of the operators $\hat{y}, \hat{p}_y, \hat{x}, \hat{p}_x$ etc. in place of the corresponding variables appearing in Equation $(L_x = yp_z - zp_y,$ cyclic,), we find that the components, $\hat{L}_x, \hat{L}_y$ and $\hat{L}_z$ of angular momentum operators satisfy the following commutation relations: $\left[\hat{L}_x, \hat{L}_y\right] = i\hbar\hat{L}_z$, (similar reaction IN cyclic order).

Time Evolution of State Vector

Postulate: You must be already familiar with time-independent treatment of the Schrodinger equation from the elementary course on quantum mechanics.

Here we postulate that time evolution of the state vector, $|\psi(t)\rangle$, is governed by the Schrodinger equation:

$$i\hbar\frac{d}{dt}|\psi(t)\rangle = \hat{H}(t)|\psi(t)\rangle$$

where $\hat{H}(t)$ is the Hamiltonian operator associated with the total energy of the system.

References

- Quantum-mechanics-physics: britannica.com, Retrieved 11, June 2020
- Photoelectric-effect: byjus.com, Retrieved 29, January 2020
- Compton-Effect, Electronic-Properties, Supplemental-Modules-(Materials-Science): eng.libretexts.org, Retrieved 13, February 2020
- Constructive-and-Destructive-Interference, youngs-double-slit-experiment: byjus.com, Retrieved 06, May 2020
- Quantrev, node19: vergil.chemistry.gatech.edu, Retrieved 25, July 2020

4

Statistical Thermodynamics and Mechanics

The branch of mechanics in which statistical methods and probability theory are applied to describe the behavior of a large number of atoms is known as statistical mechanics. A few of the topics studied under this domain are statistical thermodynamics, statistical ensemble of thermodynamics, classical mechanical formulation of statistical mechanics, etc. All these diverse principles of statistical mechanics have been carefully analyzed in this chapter.

Statistical mechanics is a branch of physics that combines the principles and procedures of statistics with the laws of both classical and quantum mechanics, particularly with respect to the field of thermodynamics. It aims to predict and explain the measurable properties of macroscopic systems on the basis of the properties and behaviour of the microscopic constituents of those systems. Statistical mechanics interprets, for example, thermal energy as the energy of atomic particles in disordered states and temperature as a quantitative measure of how energy is shared among such particles. Statistical mechanics draws heavily on the laws of probability so that it does not concentrate on the behaviour of every individual particle in a macroscopic substance but on the average behaviour of a large number of particles of the same kind.

Statistical Thermodynamics

Statistical thermodynamics is based on the fundamental assumption that all possible configurations of a given system, which satisfy the given boundary conditions such as temperature, volume and number of particles, are equally likely to occur. The overall system will therefore be in the statistically most probable configuration. The entropy of a system is defined as the logarithm of the number of possible configurations multiplied with Boltzmann's constant. While such definition does not immediately provide insight into the meaning of entropy, it does provide a straightforward analysis since the number of configurations can be calculated for any given system.

Thermodynamic Limit

Realizing the fact that macroscopic systems are made up of typically 6.023×10^{23} points to the enormity of the systems involved. Thermodynamic limit is a way to express this enormity and a natural way to express this by carrying out the limit number of particles $N \rightarrow \infty$ and volume $V \rightarrow \infty$, such that the particle density n remains fixed. Taking this limit amounts to saying that extensive properties of the system varies proportional to the size of the system i.e. N or V whereas intensive properties remain independent of the size of the system. The process can be start with a finite system with a fixed density and keep on adding more and more such identical finite systems to it such that number density remains the same causing increase only in extensive parameters without change in the intensive parameters. Not going to thermodynamic limit in the mathematical sense can lead to serious problems with the strict compliance of the extensive nature of extensive parameters such as entropy.

Macroscopic State and Microscopic State

There are two ways to characterize a macroscopic system in equilibrium under consideration. These two ways are respectively macroscopic state and microscopic state of the system. A given macroscopic state corresponds to a large number of different microstates.

- A macroscopic state also called a macrostate is described in terms of macroscopic variables such as total energy E, volume V and number of particles N. In describing a macrostate we are not at all bothered by the description of the properties of individual particles constituting the system such as their momentum, spin etc.
- A microscopic state also called microstate is specified by describing the state of each particle in a given macrostate. Counting the number of microstates in a given macrostate is the most basic task to indulge in statistical mechanics.

To further appreciate the difference, let us take a system made up of N number of non-interacting particles with total energy E. This energy is the sum of the energies E_i of individual particles, such that,

$$E = \sum_i n_i \epsilon_i$$

Where n_i is the number of particles each with energy E_i. Obviously.

$$N = \sum_i n_i$$

It may be recalled that $\in_i$ are the single particle energies which have two distinct properties:

- These are quantized, leading to quantization of total energy E.
- Their value depends on the volume to which they are confined. However, it is worth noting that as volume increases the difference between the adjacent energy levels decrease rapidly and E goes from being a discreet variable to a continuous variable.

So macrostate of a system under consideration is defined by specification of N, V and E.

This macrostate can be realized at the microscopic level in many different possible ways known as microstates. In simple words, it implies that total energy E can be distributed among N particles constituting the system in large number of different ways.

Quantum mechanically this amounts to specifying all possible independent solutions $\psi(r_1, r_2, r_3, \dots \dots \dots r_N)$ of the N particle Schrodinger equation of the system under consideration with energy eigen value E.

One of the aims of statistical mechanical formulation is count the number of all the possible microstates. In this course this number shall be denoted by Ω, which is a function of macroscopic variables of the system, which in the present case are N, V and E. Therefore, this count is a number denoted by the function $\Omega(N, V, E)$. The value $\frac{1}{\Omega}$ gives the probability of occurrence of any one of the states, which is as per the principle of equal apriori probability the same for each microstate. This $\frac{1}{\Omega}$ is also known as thermodynamic probability P_r.

The equilibrium state of the system then is the state for which function Ω(N, V, E) maximizes for each of its arguments. It is interesting to note that though a system corresponding to a given macrostate remains in a state of equilibrium, its microscopic states may keep on changing continuously.

Principle of Equal Apriori Probabilities

This principle is the back bone of the statistical mechanics for systems at equilibrium. It states that all allowed microstates compatible with a given macrostate, (i.e. subject to constraints imposed by equations $E = \sum_i n_i \epsilon_i$ and $N = \sum_i n_i$ or any other constraints on macroscopic variables involved) are equally likely.

Allowing possibility of all microstates of a given macrostate has resulted in derivation of ideal gas laws and derivation of Maxwell Boltzmann velocity distribution law which were later confirmed experimentally. Justification of this principle, therefore, lies in the results derived from it and observed experimentally. This principle is treated as one of the postulates of statistical mechanics.

Linking Thermodynamics with Statistics

We now proceed to establish a link between thermodynamics with statistics based on the state that the equilibrium state of the system then is the state for which function Ω(N, V, E) maximizes for each of its arguments leading to three conditions of equilibrium for each of the microscopic variable E, V and N. For this we consider two physical systems: system 1 and system 2 which are separately in equilibrium, described by the macroscopic variables N_1, V_1, E_1 and N_2, V_2, E_2. System 1 has $\Omega_1(N_1, V_1, E_1)$ number of microstates and system 2 has $\Omega_2(N_2, V_2, E_2)$.

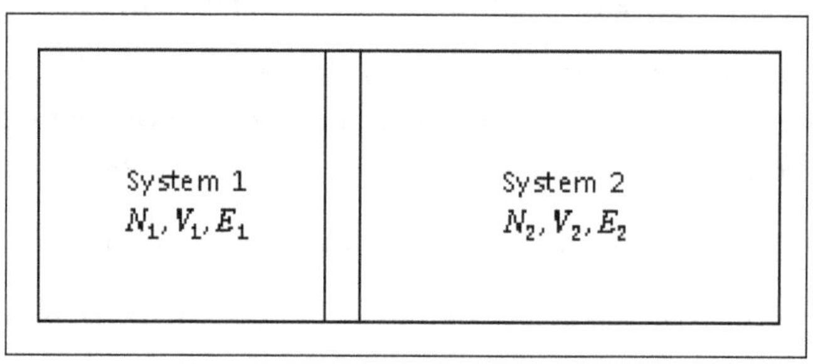

Two systems ready for making contact in three ways (a) when the partition is conducting (b) when the partition is made conducting as well as movable and (c) when the partition also allows for exchange of particles.

Case: Thermal Contact between Two Systems: Notion of Temperature in Statistical Mechanics: Let us consider first the case in which N_1, V_1 and N_2 V_2 are kept constant but energy can be exchanged by making the intervening partition conducting. However, in this case the total energy $E = E_1 + E_2$ remains conserved i.e.

$$E = E_1 + E_2 = \text{Constant}$$

At any instant system 1 is equally likely to be in any one of the microstates $\Omega_1(N_1, V_1, E_1)$ and system 2 is equally likely to be in any one of the microstates $\Omega_2(N_2, V_2, E_2)$. Then the combined system is equally likely to be any one of the microstates Ω(N, V, E), such that,

$$\Omega(N,V,E)=\Omega_1(N_1,V_1,E_1)\Omega(N_2,V_2,E_2)$$
$$=\Omega(N_1,V_1,E_1)\Omega(N_2,V_2,E-E_1)$$

Since N_1, V_1 and N_2, V_2 are constant, we can write above equation as,

$$\Omega(E)=\Omega(E_1)\Omega(E-E_1)$$

Ω number of microstates of the combined system here vary with E_1 till equilibrium is established, and this happens until Ω maximizes. The combined system then attains a macrostate which has the most number of microstates. The important point to note is that this macrostate of the system is that state in which the system spends the largest portion of its time and presents itself as an equilibrium state. Mathematical this condition when applied to above equation means,

$$\frac{\partial\Omega(E,E_1)}{\partial E_1}=0,$$

when $E_1=\overline{E_1}$, the equilibrium value of E_1 and $E_2=\overline{E_2}$.

We, therefore, obtain,

$$\left(\frac{\partial\Omega_1(E_1)}{\partial E_1}\right)_{E_1=\overline{E_1}}\Omega_2\left(\overline{E_2}\right)+\Omega_1\left(\overline{E_1}\right)\left(\frac{\partial\Omega_2(E_2)}{\partial(E_2)}\right)_{E_2=\overline{E_2}}\frac{\partial(E_2)}{\partial E_1}=0$$

Note $E_2=E-E_1$,

$$\left(\frac{\partial\Omega_1(E_1)}{\partial E_1}\right)_{E=\overline{E_1}}\Omega_2\left(\overline{E_2}\right)-\left(\frac{\partial\Omega_2(E_2)}{\partial(E_2)}\right)_{E_2=\overline{E_2}}\Omega_1\left(\overline{E_1}\right)=0$$

$$\frac{1}{\Omega_1\left(\overline{E_1}\right)}\left(\frac{\partial\Omega_1(E_1)}{\partial E_1}\right)_{E=\overline{E_1}}=\frac{1}{\Omega_2\left(\overline{E_2}\right)}\left(\frac{\partial\Omega_2(E_2)}{\partial(E_2)}\right)_{E_2=\overline{E_2}}$$

$$\left(\frac{\partial\ln\Omega_2(E_1)}{\partial E_1}\right)_{E_1=\overline{E_1}}=\left(\frac{\partial\ln\Omega_2(E_2)}{\partial E_2}\right)_{E_2=\overline{E_2}}$$

The above equation is the condition for equilibrium of the system 1 and system 2. Where,

We define $\beta=\left(\frac{\partial\ln\Omega(N,E,V)}{\partial E}\right)_{E=\overline{E},V,N}$.

So equation $\left(\frac{\partial\ln\Omega_2(E_1)}{\partial E_1}\right)_{E_1=\overline{E_1}}=\left(\frac{\partial\ln\Omega_2(E_2)}{\partial E_2}\right)_{E_2=\overline{E_2}}$ implies, $\beta_1=\beta_2$.

What have we achieved? We found that:

- When two systems are brought in contact and allowed to exchange energy through conducting partition, the exchange of energy continues till equilibrium values $E_1 = \overline{E}_1$, and $E_2 = \overline{E}_2$ of the two systems is not attained. Once this happens systems come to thermal equilibrium.

- This stage of thermal equilibrium is indicated when the respective values of the parameter β, β_1 and β_2 become equal, pointing towards zeroth law of thermodynamics which defines temperature as that parameter. It is, therefore, natural to look for this connection between β and temperature T.

- If we recall that the temperature is defined by the derivative of the entropy with respect to energy E:

$$\left(\frac{\partial S}{\partial E}\right)_{N,V} = \frac{1}{T}$$

On comparing equations $\beta = \left(\frac{\partial \ln \Omega(N,E,V)}{\partial E}\right)_{E=\overline{E},V,N}$ and $\left(\frac{\partial S}{\partial E}\right)_{N,V} = \frac{1}{T}$ we find that,

$$\frac{\Delta S}{\Delta(\ln \Omega)} = \frac{1}{\beta T} = \text{Constant}$$

Equation above was first established by Ludwig Boltzmann. He also underlined the connection between the two approaches i.e. thermodynamic approach, equation $\left(\frac{\partial S}{\partial E}\right)_{N,V} = \frac{1}{T}$ and statistical approach equation $\beta = \left(\frac{\partial \ln \Omega(N,E,V)}{\partial E}\right)_{E=\overline{E},V,N}$ and further concluded that the constant in above equation should be a universal constant, which was later named as Boltzmann constant by Planck as,

$$S = k_B \ln \Omega$$

Also,

$$\frac{1}{\beta T} = k_B \text{ or } \beta = \frac{1}{k_{BT}}$$

- According to formula $S = k_B \ln \Omega$, one can get entropy of a system by knowing the total number of microstates of a macrostate for a given system.

- If $\Omega = 1$, this corresponds to a unique completely ordered state, with entropy S = 0. As Ω increases, the number of microstates increase and more is the number of microstates more is the entropy. Therefore, in a given system disorder is manifested through the count of the microstates of a given macrostate of the system, larger this number larger is the disorder and less the degree of predictability.

- Furthermore, maximization of the number of possible microstates or probability amounts to maximizing the entropy for equilibrium.

Case: Mechanical Contact Between Two Systems: Notion of Pressure in Statistical Mechanics: If we extend the considerations of the previous case and further relax the constraint on the partition separating system 1 from system 2 by making it movable as well as conducting figure above, then the volumes of the two systems also become variables, though the total volume $V = V_1 + V_2$, remains constant and besides equation $\left(\frac{\partial \ln \Omega_2(E_1)}{\partial E_1}\right)_{E_1=\overline{E_1}} = \left(\frac{\partial \ln \Omega_2(E_2)}{\partial E_2}\right)_{E_2=\overline{E_2}}$ corresponding to thermal equilibrium,

$$\left(\frac{\partial \ln \Omega_2(E_1)}{\partial E_1}\right)_{N_1,V_1,E_1=\overline{E_1}} = \left(\frac{\partial \ln \Omega_2(E_2)}{\partial E_2}\right)_{N_2,V_2,E_2=\overline{E_2}}$$

We have another equation which can be obtained in an analogous way as in previous case giving another condition for the system to be in mechanical equilibrium as well,

$$\left(\frac{\partial \ln \Omega_2(V_1)}{\partial V_1}\right)_{N_1,E_1,V_1=\overline{V_1}} = \left(\frac{\partial \ln \Omega_2(V_2)}{\partial V_2}\right)_{N_2,E_2,V_2=\overline{V_2}}$$

And we can define another parameter η,

$$\eta = \left(\frac{\partial \ln \Omega(N,E,V)}{\partial V}\right)_{V=\overline{V},E,N}$$

On comparing with, again recalling thermodynamic formula,

$$\left(\frac{\partial S}{\partial V}\right)_{N,E} = \frac{P}{T}$$

Equation $\left(\frac{\partial \ln \Omega_2(V_1)}{\partial V_1}\right)_{N_1,E_1,V_1=\overline{V_1}} = \left(\frac{\partial \ln \Omega_2(V_2)}{\partial V_2}\right)_{N_2,E_2,V_2=\overline{V_2}}$ implies that,

$$\frac{\Delta S}{\Delta(\ln \Omega)} = \frac{P}{\eta T} = \text{Constant}$$

Therefore, we can conclude:

The equilibrium condition during thermal and mechanical contacts, such that $\frac{P}{\eta T} = k_B$ or $\eta = \frac{P}{K_B T}$ becomes:

$$\frac{P_1}{T_1} = \frac{P_2}{T_2}$$

i.e. $T_1 = T_2$ and $P_1 = P_2$.

Case: Chemical Contact between Two Systems: Notion of Chemical Potential in Statistical Mechanics: Next if we consider besides thermal contact, partition such that it not only allows thermal contact but also allows exchange of particles between system 1 and system 2, but keep the volume fixed. Then only total Number of particles $N = N_1 + N_2$ and total Energy $E = E_1 + E_2$ remain fixed but V_1, V_2, E_1 and E_2 of the two systems can change. So besides equation,

$$\left(\frac{\partial \ln \Omega_2(E_1)}{\partial E_1}\right)_{E_1=\overline{E_1}} = \left(\frac{\partial \ln \Omega_2(E_2)}{\partial E_2}\right)_{E_2=\overline{E_2}}$$

$$\left(\frac{\partial \ln \Omega_1(E_1)}{\partial E_1}\right)_{N_1,V_1,E_1=\overline{E_1}} = \left(\frac{\partial \ln \Omega_2(E_2)}{\partial E_2}\right)_{N_2,V_2,E_2=\overline{E_2}}$$

We have another equation which can be obtained in an analogous way as in case I giving another condition for the system to be in chemical equilibrium as well,

$$\left(\frac{\partial \ln \Omega_1(N_1)}{\partial N_1}\right)_{N_1=\overline{N_1},E_1,V_1} = \left(\frac{\partial \ln \Omega_2(N_2)}{\partial N_2}\right)_{N_2=\overline{N_2},E_2,V_2}$$

And we can define another parameter μ,

$$\mu = \left(\frac{\partial \ln \Omega(N,E,V)}{\partial N}\right)_{N=\overline{N},E,V}$$

On comparing with, again recalling thermodynamic formula,

$$\left(\frac{\partial S}{\partial N}\right)_{N,E,V} = -\frac{\mu}{T}$$

We get,

$$\frac{\Delta S}{\Delta(\ln \Omega)} = -\frac{\mu}{T} = \text{Constant}$$

Therefore,

- The equilibrium condition during Chemical contact, such that $-\frac{\mu}{T} = k_B$ or $\mu = -k_B T$ becomes,

  $$\frac{\mu_1}{T_1} = \frac{\mu_2}{T_2}$$

  i.e. $T_1 = T_2$ and $\mu_1 = \mu_2$.

- If we combine the three case discussed above such that if we allow exchange of energy, change of volume and exchange of particles simultaneously, the conditions of equilibrium are:

  $T_1 = T_2$, $P_1 = P_2$ and $\mu_1 = \mu_2$.

Recipe for Deriving Thermodynamics from Statistical Route

Now we are ready to derive thermodynamics from statistical considerations. Recipe is as given below:

- For the macrostate (N, V, E), of a given system count the number of all possible microstates accessible to the macrostate denoted by $\Omega(N, V, E)$.
- The entropy of the system can be calculated by equation the equation $S = k_B \ln \Omega$.
- The intensive parameters, T, P and μ can then be calculated by the following thermodynamic relations,

$$\frac{1}{\ } \left(\frac{\partial S}{\partial}\right)_{N,V}$$

$$\frac{P}{T} = \left(\frac{\partial S}{\partial V}\right)_{N,E}$$

$$\frac{\mu}{T} = -\left(\frac{\partial S}{\partial N}\right)_{V,E}$$

Or alternatively,

$$P = \frac{\left(\frac{\partial S}{\partial V}\right)_{N,E}}{\left(\frac{\partial S}{\partial E}\right)_{N,V}} = -\left(\frac{\partial E}{\partial V}\right)_{N,S}$$

And

$$\mu = -\frac{\left(\frac{\partial S}{\partial N}\right)_{V,E}}{\left(\frac{\partial S}{\partial E}\right)_{N,V}} = \left(\frac{\partial E}{\partial N}\right)_{V,S}$$

With $U \equiv E$, rest of the thermo dynamics follows from:

- Helmholtz free energy: $F = E - Ts$
- Gibbs free energy: $G = E - TS + PV$
- Enthalpy: $H = E + PV = G + TS$

Furthermore, specific heat at constant volume, C_V and Specific heat at constant pressure, C_P are given by,

$$C_V = \left(\frac{\partial E}{\partial T}\right)_{N,V}$$

$$C_P = \left(\frac{\partial H}{\partial T}\right)_{N,P}$$

In nut shell, once one knows the entropy S(N, E, V) of the system, equilibrium behavior of the system can be known by using thermodynamics.

Statistical Ensemble of Thermodynamics

Micro-Canonical Ensemble

The micro-canonical ensemble is defined as a collection of systems with exactly the same number of particles and with the same volume micro canonical ensemble, representing a system having constant energy E, constant number of particles N and constant volume V. It describes a system which is completely isolated system from its surroundings, a highly idealized situation seldom encountered in real situations. Within these constraints particles of the system can be in any one of the allowed microstates with equal apriori probability.

Phase Space and Number of Microscopic States

As yet, till this point we have described the enumeration of microstates in terms of calculating the volume of the allowed phase space. This enumeration of microstates is possible through Heisenberg's Uncertainty Principle according to which phase space can be regarded as made up of elementary cells of the minimum volume h^d where d is the dimension of the phase space representing single particle quantum state. In the case of an N particle system with phase space having 6N dimensions, d = 3N.

Micro-Canonical Ensemble Distribution Function

To define micro-canonical distribution function we must ponder on a crucial question, can we have an absolutely isolated system, because in actual systems, there is always some contact with the surroundings and system does not have a sharp value and it will have a width Δ around E. Hence system has a range of energy values $\left(E-\frac{1}{2}\Delta, E+\frac{1}{2}\Delta\right)$ where Δ is small compared to E. Doing this implies that a system has a large number of possible microstates in the phase space lying anywhere between a hyper-shell are equally apriori probable such that,

$$E-\frac{1}{2}\Delta \le H \le E+\frac{1}{2}\Delta$$

This defines a region in phase space having volume,

$$\omega = \int \Theta\left(E+\frac{1}{2}\Delta-H\right)\Theta\left(H-E+\frac{1}{2}\Delta\right)d^{3N}q\,d^{3Np}$$

Where Θ represents a unit step function defined by,

$$\Theta(x) = \begin{cases} 1 & \text{if } x \geq 0 \\ 0 & \text{otherwise} \end{cases}$$

The step functions sitting as integrand restrict the region of integration defined by equation $E - \frac{1}{2}\Delta \leq H \leq E + \frac{1}{2}\Delta$. It is obvious that allowed phase space volume is a function of E, N, V and Δ. Then a microcanonical ensemble is a collection of systems, for which probability distribution function ρ(q, p) in equilibrium is given by,

$$\rho(q,p) = \begin{cases} \text{Constant for } E - \frac{1}{2}\Delta \leq H \leq E + \frac{1}{2}\Delta \\ 0 \qquad\qquad \text{otherwise} \end{cases}$$

Now if f(q, p) is a physical quantity then the microcanonical ensemble average is given by,

$$\langle f \rangle = \frac{\int \rho(q,p) f(p,q) d^{3N}q\, d^{3Np}}{\int \rho(q,p) d^{3N}q\, d^{3Np}} = \frac{\int f(p,q) d^{3N}q\, d^{3Np}}{\int d^{3N}q\, d^{3Np}}$$

Canonical Ensemble

In physics, Canonical ensemble is a functional relationship for a system of particles that is useful for calculating the overall statistical and thermodynamic behaviour of the system without explicit reference to the detailed behaviour of particles. The canonical ensemble was introduced by J. Willard Gibbs, a U.S. physicist, to avoid the problems arising from incompleteness of the available observational data concerning the detailed behaviour of a system of interacting particles—for example, molecules in a gas.

One way to describe a system of particles is to state explicitly the position and momentum (i.e., mass times velocity) of each particle. If there are N particles and each particle has s modes in which it can move, 2sN values are required to specify its state. This system can then be described as a point in a 2sN-dimensional space (called gamma [Γ] space). As time passes, changes in the details of the system would correspond to movement of the point in the Γ space. An ensemble is a large number of similar systems, as described by a collection of points in Γ space.

A canonical ensemble (or, more explicitly, macrocanonical ensemble) is an ensemble for which the density of points in Γ space varies exponentially with the total energy E of the system: $\rho = Ae^{-E/\theta}$, in which A and theta (θ) are constants of the system. If the system is in equilibrium at absolute temperature T, its gross (macroscopic) behaviour will be described by taking the average behaviour of a system in a canonical ensemble in which θ = kT. The constant k is called Boltzmann's constant.

A microcanonical ensemble consists of systems all of which have the same energy and is often found useful in describing isolated systems in which the total energy is a constant. Such macrocanonical and microcanonical ensembles are examples of petit ensembles, in that the total number of particles in the system is specified.

A grand ensemble is any ensemble for which the restriction of a constant number of particles is abandoned. Such a description is more general and is particularly applicable to systems in which the number of particles varies, e.g., chemically reacting systems.

Grand Canonical Ensemble

The grand canonical ensemble is a generalization of the canonical ensemble where the restriction to a definite number of particles is removed. This is a realistic representation when then the total number of particles in a macroscopic system cannot be fixed.

Grand Canonical Distribution Function: System and Heat Bath Approach

We follow the same approach followed for canonical ensemble system and heat bath approach. We take our system (designated A) and put it in a heat bath (designated A') such that system and the heat bath (designated $A_T = A + A'$), are allowed to now exchange both energy and particles, keeping system and heat bath in equilibrium at a common temperature T and common chemical potential μ. Heat bath is very large having a large heat capacity. The composite system (A_T) together is completely isolated, such that if at any time if system A has energy E_r and number of particles N_r and heat bath A' having energy E_r' and number of particles N_r' then the total energy E_T and total number of particles N_T are constant i.e.

$$E_s + E_s = E_T = \text{Constant}$$

$$N_r + N_r' = N_T = \text{Constant}$$

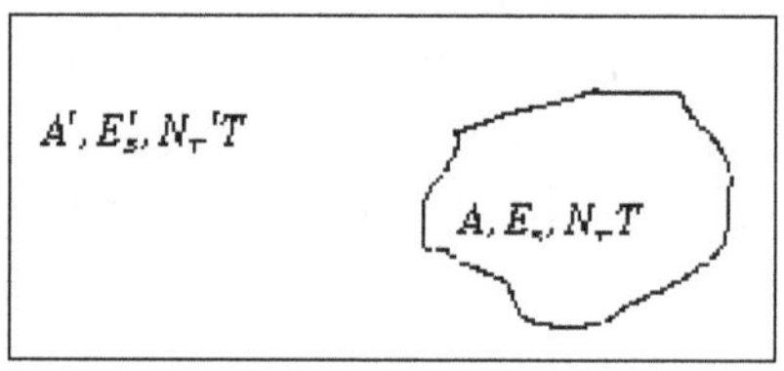

Heat bath A' at equilibrium with system A at temperature T and chemical potential μ.

Since heat bath is very large, $E_s << E_T$ and $N_s << N_T$ i.e

$$\frac{E_s}{E_T} = \left(1 - \frac{E_s'}{E_T}\right) << 1$$

$$\frac{N_r}{N_T} = \left(1 - \frac{N_r'}{N_T}\right) << 1$$

Since the heat bath is very large the number of states compatible to it with energy E_s' and particles $N_r', \Omega'(E_s', N_r')$ is also very large independent of the nature of the reservoir. It is worth noting that larger the number of states available to the reservoir larger is the probability of it assuming energy E_s' and having particles N_r' and consequently the system immersed in heat bath assuming energy

E_s and having number of particles N_r. As per the principle of equal apriori probability, different microstates in the reservoir are equally likely to occur, the probability P_r of system A, occurring with energy E_s and number of particles N_r, is proportional to the number of microstates which the reservoir has i.e. $\Omega'(E_s', N_r')$. Therefore,

$$P_{s,r}\,\alpha\,\Omega'(E_s', N_r') \equiv \Omega'(E_T - E_s, N_T - N_r)$$

Noting the inequalities in equations $\frac{E_s}{E_T} = \left(1 - \frac{E_s'}{E_T}\right) << 1$ and $\frac{N_r}{N_T} = \left(1 - \frac{N_r'}{N_T}\right) << 1$ we can expand the right hand side around $E_s' = E_T$ and $N_r' = N_T$, which means around $E_s = 0$ and $N_r = 0$ Recalling that Ω' is a very large number which varies very rapidly, it is therefore, more convenient to deal with logarithm of this function which is a slowly varying function. So,

$$\ln\Omega'(E_s', N_r') = \ln\Omega'(E_T, N_T) + \left(\frac{\partial \ln\Omega'}{\partial E'}\right)_{E_r'=E_T}(E_s' - E_T)$$
$$+\left(\frac{\partial \ln\Omega'}{\partial N'}\right)_{E_s'=E_T}(N_r' - N_T) + \cdots$$

Or

$$\ln\Omega'(E_r', N_r') \approx \ln\Omega'(E_T, N_T) + \left(\frac{\partial \ln\Omega'}{\partial E'}\right)_{E_s'=E_T}(-E_s)$$
$$+\left(\frac{\partial \ln\Omega'}{\partial N'}\right)_{N_r'=N_T}(-N_r)\cdots$$

Recalling $\left(\frac{\partial \ln\Omega'}{\partial E'}\right) = \beta' = \frac{1}{k_B T'}$ and at equilibrium $\beta' = \frac{1}{k_B T'} = \beta = \frac{1}{k_B T}, \left(\frac{\partial \ln\Omega'}{\partial N'}\right) = \alpha' = -\frac{\mu'}{k_B T'} = \alpha = -\frac{\mu}{k_B T}$.

Equation, $\ln\Omega'(E_s', N_r') = \ln\Omega'(E_T, N_T) + \left(\frac{\partial \ln\Omega'}{\partial E'}\right)_{E_r'=E_T}(E_s' - E_T) + \left(\frac{\partial \ln\Omega'}{\partial N'}\right)_{E_s'=E_T}(N_r' - N_T) + \cdots$ can be written as,

$$\ln\Omega'(E_s', N_r') \approx \text{constant} - \beta E_s - \alpha N_r$$

Or

$$\Omega'(E_s', N_r') \approx \text{Constant} - e^{\beta E_s - \alpha N_r}$$

From equations $\frac{N_r}{N_T} = \left(1 - \frac{N_r'}{N_T}\right) << 1$ and above equation we get,

$$P_{s,r}\ \alpha - e^{-\beta(E_s-\mu N_r)}$$

Since sum of all probabilities must sum to 1, normalising, () we get,

$$P_{s,r} = \frac{e^{-\beta(E_s-\mu N_r)}}{\sum_{s,r} e^{-\beta(E_s-\mu N_r)}}$$

Where, summation is over all the (E_s, N_r) states which system A can take. Equation above is the Grand canonical distribution function which provides the probability of the system in state with energy E_s and number of particles N_r.

Furthermore, since above equation has been derived by using very fundamental ideas of physics involving conservation of energy, extensiveness of energy $E_T = E_r + E_r^{'}$, conservation of number of particles, extensiveness of number of particles and slow and smooth variation of $\ln\Omega(E,N)$, independent of the choice of quantum mechanics or classical mechanics.

Grand Canonical Distribution Function: Most Probable Value Approach

We consider an ensemble of $\mathbb{N}$ identical systems which are allowed to mutually share total energy $\mathbb{N} < E >$ and a total number of particles $\mathbb{N} < N >$. Let n_{sr} denote the number of system which has at any instant t an amount of energy E_s and number of particles N_r then these together satisfy the following three conditions:

$$\sum_{s,r} n_{sr} = \mathbb{N}$$

$$\sum_{r} n_{sr}E_s = \mathbb{N} < E >$$

$$\sum_{r} n_{sr}N_r = \mathbb{N} < N >$$

Any set of values $\{n_{sr}\}$ which satisfy conditions stated in above three equations is a possible way of distributing energy and particles among the members of the ensemble. Number of such possible ways $W\{n_{rs}\}$, of distributing energy and particles among the members of the ensemble is given by,

$$\mathbb{W}\{n_{sr}\} = \frac{\mathbb{N}!}{\prod_{n_{sr}}(n_{sr}!)}$$

According to principle of equal apriori probability, all possible distribution of energy and particles among the members of the ensemble are equally likely, the frequency of achieving a given set $\{n_{rs}\}$ is directly proportional to the number $W\{n_{rs}\}$. Out of all these sets the most probable set shall be the one for which $W\{n_{rs}\}$ is maximum. Once we have identified this set say $\{n_{rs}^{*}\}$, we shall be physically interested in this distribution. However statistically we cannot ignore other $\{n_{rs}\}$'s with their corresponding weights or frequencies to calculate the expectation values or mean values of the numbers n_r as given below:

$$\langle n_{rs} \rangle = \frac{\sum'_{\{n_{sr}\}} n_r \mathbb{W}\{n_{sr}\}}{\sum'_{\{n_{sr}\}} \mathbb{W}\{n_{sr}\}}$$

Significance of n^*_{sr} lies in the fact that the fraction $\frac{n^*_{sr}}{\mathbb{N}}$ is the grand canonical probability P_{sr}.

Let us now calculate the desirable numbers $\{n^*_{sr}\}$ and $\langle n_{sr} \rangle$. Again for the reasons stated earlier $\ln \mathbb{W}$ is a slowly varying function and we shall use it to proceed further, which implies,

$$\ln \mathbb{W} = \ln \mathbb{N}! - \sum_{sr} \ln n_{sr}!$$

In thermodynamic limit apply Stirlings formula and equation $\sum_r n_{sr} E_s = \mathbb{N} < E >$ becomes,

$$\ln \mathbb{W} = \mathbb{N} \ln \mathbb{N} - \mathbb{N} - \sum_{sr} n_{sr} \ln n_{sr} + \sum_{sr} n_{sr}$$

$$= \mathbb{N} \ln \mathbb{N} - \sum_r n_{sr} \ln n_{sr}$$

If n_{sr} changes by a small amount $\ln \mathbb{W}$ also changes by a small amount as given below:

$$\delta(\ln \mathbb{W}) = \sum_r (\ln n_{sr} + 1)\delta n_{sr}$$

Now for n_{sr} to be maximum $\delta(\ln \mathbb{W})$ should become zero, provided equations $\sum_{s,r} n_{sr} = \mathbb{N}$ and $\sum_r n_{sr} E_s = \mathbb{N} < E >$ and $\sum_r n_{sr} N_r = \mathbb{N} < N >$ also satisfy the following conditions simultaneously,

$$\sum_r \delta n_{sr} = 0$$

$$\sum_{sr} E_s \delta n_{sr} = 0$$

$$\sum_r N_r \delta n_{sr} = 0$$

Now to get the set $\{n^*_r\}$, method of Lagrange's undetermined multiplier can be used according to which equations $\sum_r \delta n_{sr} = 0$ and $\sum_{sr} E_s \delta n_{sr} = 0$, together yield the condition,

$$\sum_r \left(-(\ln n_{sr} + 1) - \alpha N_r - \beta E_s - \gamma\right)\delta n_{sr} = 0$$

Where α, β and γ are Lagrange's undetermined multipliers. Since δn_r is completely arbitrary, for equation $\ln \mathbb{W} = \mathbb{N}\ln\mathbb{N} - \mathbb{N} - \sum_{sr} n_{sr}\ln n_{sr} + \sum_{sr} n_{sr} = \mathbb{N}\ln\mathbb{N} - \sum_{r} n_{sr}\ln n_{sr}$ to be satisfied its coefficients must become zero, i.e. for all r,

$$\left(-\left(\ln n_r^* + 1\right) - \alpha N_r - \beta E_s - \gamma\right) = 0$$

Or

$$\ln n_{sr}^* = -1 - \alpha N_r - \beta E_s - \gamma$$

Or

$$n_{sr}^* = e^{-(1+\gamma)} e^{-\beta E_r - \alpha N_r} = C e^{-\beta E_r - \alpha N_r}$$

Where now C, α and β are two undetermined parameters and above equation gives us the most probable distribution. To calculate C, we note,

$$\sum_r n_{sr}^* = \mathbb{N} = C\sum_{sr} e^{-\beta E_s - \alpha N_r}$$

Or

$$C = \frac{\mathbb{N}}{\sum_{sr} e^{-\beta E_s - \alpha N_r}}$$

Therefore,

$$P_{sr} = \frac{n_{sr}^*}{\mathbb{N}} = \frac{e^{-\beta E_s - \alpha N_r}}{\sum_{sr} e^{-\beta E_s - \alpha N_r}}$$

In Thermodynamic limit at equilibrium $\frac{n_{sr}^*}{\mathbb{N}} \approx \frac{\langle n_{sr}\rangle}{\mathbb{N}}$, therefore,

$$\frac{\langle n_{sr}\rangle}{\mathbb{N}} \approx \frac{n_{sr}^*}{\mathbb{N}} = \frac{e^{-\beta E_s - \alpha N_r}}{\sum_{sr} e^{-\beta E_s - \alpha N_r}}$$

The undetermined parameters α and β are determined by the equations for average number of particles $\langle N\rangle$ and average energy $\langle E\rangle$ in each system of the ensemble with prior knowledge of their values:

$$\langle N\rangle = \frac{\sum_{sr} N_r e^{-\beta E_s - \alpha N_r}}{\sum_{sr} e^{-\beta E_s - \alpha N_r}} = -\frac{\partial}{\partial \alpha}\left\{\ln \sum_{sr} N_r e^{-\beta E_s - \alpha N_r}\right\}$$

$$\langle E\rangle = \frac{\sum_{sr} E_s e^{-\beta s - \alpha N_r}}{\sum_{sr} e^{-\beta E_s - \alpha N_r}} = -\frac{\partial}{\partial \beta}\left\{\ln \sum_{sr} E_r e^{-\beta E_s - \alpha N_r}\right\}$$

Link between Grand Canonical Ensemble and Thermodynamics: Thermodynamic Potential

To establish a link between the Grand canonical ensemble and thermodynamics, we introduce a quantity q, called q potential as follows:

$$q = \ln\left\{\sum_{sr} e^{-\beta E_s - \alpha N_r}\right\}$$

q is a function of α, β and all the energy states E_s. Taking the differential of q, we have,

$$dq = -\frac{\sum_{sr} N_r e^{-\beta E_s - \alpha N_r}}{\sum_{sr} e^{-\beta E_s - \alpha N_r}} d\alpha - \frac{\sum_{sr} E_s e^{-\beta E_s - \alpha N_r}}{\sum_{sr} e^{-\beta E_s - \alpha N_r}} d\beta$$

$$-\beta \sum_{sr} \frac{e^{-\beta E_s - \alpha N_r}}{\sum_{sr} e^{-\beta E_s - \alpha N_r}} dE_s$$

Or using equations $\frac{\langle n_{sr}\rangle}{N} \approx \frac{n^*_{sr}}{N} = \frac{e^{-\beta E_s - \alpha N_r}}{\sum_{sr} e^{-\beta E_s - \alpha N_r}}$, $\langle N\rangle = \frac{\sum_{sr} N_r e^{-\beta E_s - \alpha N_r}}{\sum_{sr} e^{-\beta E_s - \alpha N_r}} = -\frac{\partial}{\partial \alpha}\left\{\ln \sum_{sr} N_r e^{-\beta E_s - \alpha N_r}\right\}$ and $\langle E\rangle = \frac{\sum_{sr} E_s e^{-\beta s - \alpha N_r}}{\sum_{sr} e^{-\beta E_s - \alpha N_r}} = -\frac{\partial}{\partial \beta}\left\{\ln \sum_{sr} E_r e^{-\beta E_s - \alpha N_r}\right\}$, we have,

$$dq = -\langle N\rangle d\alpha - \langle E\rangle d\beta - \frac{\beta}{N}\sum_{sr}\langle n_{sr}\rangle dE_s$$

Which can be further written as,

$$d\left(q + \alpha\langle N\rangle + \beta\langle E\rangle\right) = \beta\left(-\frac{\alpha}{\beta} d\langle N\rangle - d\langle E\rangle - \frac{1}{N}\sum_{sr}\langle n_{sr}\rangle dE_s\right)$$

Now if we compare the expression with in the parenthesis with the first law of thermodynamics:

$$\delta Q = -\frac{\alpha}{\beta} d\langle N\rangle - d\langle E\rangle + \delta W$$

We can arrive at the following correspondence,

$$\mu = -\frac{\alpha}{\beta}$$

$$\delta W = -\frac{1}{N}\sum_{sr}\langle n_{sr}\rangle dE_s$$

So that,

$$d\left(q + \alpha\langle N\rangle + \beta\langle E\rangle\right) = \beta\delta Q$$

A closer look at above equation and comparing it with $dS=\frac{\delta Q}{T}$ tells us that the coefficient β on the right hand side should correspond to reciprocal of temperature T, so that,

$$\beta=\frac{1}{k_B T}$$

From equation $\mu=-\frac{\alpha}{\beta}$, it immediately follows that,

$$\alpha=-\frac{\mu}{k_B T}$$

Left hand side of equation $d\left(q+\alpha\langle N\rangle+\beta\langle E\rangle\right)=\beta\delta Q$, then provides correspondence of the quantity $q+\alpha\langle N\rangle+\beta\langle E\rangle$ with entropy, such that,

$$\frac{S}{k_B}=q+\alpha\langle N\rangle+\beta\langle E\rangle$$

Or

$$q=\frac{S}{k_B}-\alpha\langle N\rangle-\beta\langle E\rangle=\frac{S}{k_B}-\frac{\mu\langle N\rangle}{k_B T}-\frac{\langle E\rangle}{k_B T}=\frac{TS+\mu\langle N\rangle-\langle E\rangle}{k_B T}$$

The function $k_B Tq=TS+\mu\langle N\rangle-\langle E\rangle$ is the well-known Thermodynamic Potential Φ which describes an open, isochoric and isothermal collection,

$$\Phi(T,\mu,V)=TS+\mu\langle N\rangle-\langle E\rangle=qk_B T$$

Recalling that Gibbs free energy $G=\mu\langle N\rangle=\langle E\rangle-TS+PV$, equation $q=\frac{S}{k_B}-\alpha\langle N\rangle-\beta\langle E\rangle=\frac{S}{k_B}-\frac{\mu\langle N\rangle}{k_B T}-\frac{\langle E\rangle}{k_B T}=\frac{TS+\mu\langle N\rangle-\langle E\rangle}{k_B T}$ becomes,

$$q=\ln\left\{\sum_{sr}{}^{-\beta E_s-\alpha N_r}\right\}=\frac{PV}{k\ T}$$

Equation above provides a bridge between thermodynamics and grand canonical ensemble

Grand Canonical Partition Function and Thermodynamic Functions

We can define grand canonical partition function $\mathbb{Z}$,

$$\mathbb{Z}=\left\{\sum_{sr}e^{-\frac{E_s}{k_B T}+\frac{\mu N_r}{k_B T}}\right\}$$

Defining $e^{\frac{\mu}{k_BT}} = f$ called fugacity above equation can be written as,

$$\mathbb{Z} = \left\{ \sum_{sr} (f)^{N_r} e^{-\frac{E_s}{k_BT}} \right\} = \left\{ \sum_{r} (f)^{N_r} \left(\sum_{s} e^{-\frac{E_s}{k_BT}} \right) \right\}$$

The expression $Z_{N_r}(V,T) = \left(\sum_s e^{-\frac{E_s}{k_BT}} \right)$ is the canonical partition function for a fixed N_r and has in it via energy, dependence on volume. Therefore, Grand canonical partition function can be written as,

$$\mathbb{Z} = \left\{ \sum_{N_r=0}^{\infty} (f)^{N_r} Z_{N_r}(V,T) \right\}$$

Hence q is related to the partition function via the relation,

$$q = \ln \mathbb{Z}(f,V,T) = \frac{PV}{k_BT}$$

Now we are in a position to write down the thermodynamic functions of interest in terms of q or $\mathbb{Z}$.

Pressure:

$$P = \frac{k_BT}{V} q(f,V,T) = \ln \mathbb{Z}(f,V,T)$$

Average number of particles in equilibrium:

$$N = \langle N \rangle = k_BT \left[\frac{\partial q(\mu,V,T)}{\partial \mu} \right]_{V,T} = \left[f \frac{\partial q(f,V,T)}{\partial f} \right]_{V,T}$$

Internal Energy:

$$E = \langle E \rangle = -\left[\frac{\partial q(f,V,T)}{\partial \beta} \right]_{V,f} = k_BT^2 \left[\frac{\partial q(f,V,T)}{\partial T} \right]_{V,f}$$

One can get E as a function of N, V, T by eliminating f between above two equations. Using this functional form, taking derivative with respect to T gives specific heat at constant volume.

Equation of state: It is a relationship between P, V, T and can be obtained by eliminating *f* between equations $P = \frac{k_BT}{V} q(f,V,T) = \ln \mathbb{Z}(f,V,T)$ and $N = \langle N \rangle = k_BT \left[\frac{\partial q(\mu,V,T)}{\partial \mu} \right]_{V,T} = \left[f \frac{\partial q(f,V,T)}{\partial f} \right]_{V,T}$.

Helmholtz free energy:

$$F = E - TS + PV - PV = G - PV = N\mu - PV$$

$$= Nk_BT \ln f - k_BT \ln \mathbb{Z}(f,V,T) = -k_BT \ln \frac{\mathbb{Z}(f,V,T)}{f^N}$$

Interestingly, argument of the logarithm is canonical distribution function.

Entropy:

$$S=\frac{E-F}{T}=k_BT\left[\frac{\partial q(f,V,T)}{\partial T}\right]-Nk_B\ln f+k_Bq(f,V,T)$$

Grand Canonical Partition Function and Single Orbital Partition Functions

Keeping equation $\sum_r n_{sr}E_s=\mathbb{N}<E>$ in mind the grand partition function $\mathbb{Z}=\left\{\sum_{N_r=o}^{\infty}(f)^{N_r}Z_{N_r}(V,T)\right\}$,

$$\mathbb{Z}==\left\{\sum_{N_r=o}^{\infty}(f)^{N_r}Z_{N_r}(V,T)\right\}$$

By recalling that,

$$Z_{N_r}(V,T)=\left(\sum_s e^{-\frac{E_s}{k_BT}}\right)=\left(\sum_{\{N_r\}} e^{-\Sigma_r\frac{N_rE_r}{k_BT}}\right)$$

With $\sum_{\{N_r\}}N_r=N$, can be written as,

$$\mathbb{Z}==\left\{\sum_{N_r=o}^{\infty}(f)^{N_r}\sum_{\{N_r\}}\left(e^{-\Sigma_r\frac{N_rE_r}{k_BT}}\right)\right\}$$

Or

$$\mathbb{Z}==\left\{\sum_{N_r=o}^{\infty}\sum_{\{N_r\}}(f)^{\Sigma_r N_r}\left(e^{-\Sigma_r\frac{N_rE_r}{k_BT}}\right)\right\}$$

Or

$$\mathbb{Z}=\left\{\sum_{N_r=o}^{\infty}\sum_{\{N_r\}}\prod_r\left(fe^{-\frac{E_r}{k_BT}}\right)^{N_r}\right\}$$

This can be further written as,

$$\mathbb{Z}=\left\{\prod_r\sum_{N_r=o}^{\infty}\left(fe^{-\frac{E_r}{k_BT}}\right)^{N_r}\right\}$$

Here the product is over the partition function for a single orbital only with all possible number of particles. In this form grand particle function becomes a very useful result providing us to find the grand partition function from one orbital partition function. We shall see its utility later in deriving Fermi-Dirac and Bose-Einstein distribution function which is decided the number of particles allowed in an orbital or state.

Here we discuss two applications of Grand canonical ensemble. First one is our prototype example of an ideal gas and second one is to derive Fermi-Dirac and Bose-Einstein distribution function. The real effectiveness of grand canonical ensemble becomes useful when one applies it to systems involving inter particle interaction and systems having quantum statistical effect.

Classical Ideal Gas

The grand partition function for a macroscopic system is given by,

$$\mathbb{Z}==\left\{\sum_{N=o}^{\infty}(f)^{N} Z_{N}(V,T)\right\}$$

Where $Z_N(V, T)$ is the canonical partition function and $f = e^{\frac{\mu}{k_B T}}$ is fugacity. For an ideal gas the partition function of such a system can be written as made up single particle partition function $Z_1(V, T)$:

$$Z_N(V,T)=\frac{\left[Z_1(V,T)\right]^N}{N!}$$

Here factor N! in the denominator takes care of in distinguishability of particles. Once again recalling that since particles in the system are free to be anywhere in the volume in which they are enclosed, therefore, $Z_1(V, T)$ shall be proportional to V and a function of T, so that

$$Z_1(V,T)=Vg(T)$$

Where g(T) is a function of T alone. The classical ideal gas in canonical ensemble $g(T)=\left(\frac{2\pi m k_B T}{h^2}\right)^{3/2}$

Thus the grand partition function of the ideal gas can be written as,

$$\mathbb{Z}=\sum_{N=o}^{\infty}\frac{\left[fVg(T)\right]^N}{N!}=e^{fVg(T)}$$

Hence the q potential for the ideal gas can be written as,

$$q(f,V,T)=\ln\mathbb{Z}(f,V,T)=fVg(T)$$

Pressure: From equation $P=\frac{k_B T}{V}q(f,V,T)=\ln\mathbb{Z}(f,V,T)$ we have Pressure P,

$$P=\frac{k_B T}{V}q(f,V,T)=k_B T f g(T)$$

Average number of particles: From equation $N=\langle N\rangle=k_B T\left[\frac{\partial q(\mu,V,T)}{\partial\mu}\right]_{V,T}=\left[f\frac{\partial q(f,V,T)}{\partial f}\right]_{V,T}$ we have average number of particles N,

$$N = \left[f \frac{\partial q(f,V,T)}{\partial f} \right]_{V,T} = fVg(T)$$

Internal Energy, Helmholtz Free Energy, Entropy: From equation $E = \langle E \rangle = -\left[\frac{\partial q(f,V,T)}{\partial \beta} \right]_{V,f}$

$E = \langle E \rangle = -\left[\frac{\partial q(f,V,T)}{\partial \beta} \right]_{V,f}$ we have the Internal Energy E,

$$E = fVk_B T^2 \left[\frac{\partial g(T)}{\partial T} \right]_{V,f}$$

From equation $F = E - TS ... -k_B T \ln \frac{\mathbb{Z}(f,V,T)}{f^N}$ we have Helmholtz free energy,

$$F = Nk_B T \ln f - k_B T \ln \mathbb{Z}(f,V,T) = Nk_B T \ln f - k_B T fVg(T)$$

And finally from equation $S = \frac{E-F}{T} = k_B T \left[\frac{\partial q(f,V,T)}{\partial T} \right] - Nk_B \ln f + k_B q(f,V,T)$ we have entropy,

$$S = k_B fV \left[T \left[\frac{\partial g(T)}{\partial T} \right] + g(T) \right] - Nk_B \ln f$$

Equation of State: Eliminating f from equation $q(f,V,T) = \ln \mathbb{Z}(f,V,T) = fV g(T)$ and $P = \frac{k_B T}{V} q(f,V,T) = k_B T f g(T)$ we get,

$$PV = Nk_B T$$

Equation of state is independent of the functional form of g(T).

Internal energy: Eliminating f between equation $N = \left[f \frac{\partial q(f,V,T)}{\partial f} \right]_{V,T} = fVg(T)$ and $E = fVk_B T^2 \left[\frac{\partial g(T)}{\partial T} \right]_{V,f}$.

$$E = fVk_B T^2 \left[\frac{\partial g(T)}{\partial T} \right]_{V,f} = \frac{Nk_B T^2}{g(T)} \left[\frac{\partial g(T)}{\partial T} \right]_{V,f}$$

Specific heat at constant volume: Taking the derivative of above equation with respect to temperature T, specific heat at constant volume is given by,

$$C_V = \frac{Nk_B \left[-T^2 \left[\frac{\partial g(T)}{\partial T} \right]_{V,f}^2 + g(T) T^2 \frac{\partial^2 g(T)}{\partial T^2} + 2T g(T) \frac{\partial g(T)}{\partial T} \right]}{\left[g(T) \right]^2}$$

To see whether E and CV indeed are well known results for classical ideal gas, we require explicit temperature dependence of (T) which is known to be proportional to some power of T. If we suppose this to be g(T) α T^n we get,

$$E = \frac{Nk_BT^2}{g(T)}\left[\frac{\partial g(T)}{\partial T}\right]_{V,f} = n\frac{Nk_BT^2}{T^n}T^{n-1} = nNk_BT$$

For an ideal gas n=3/2 which gives us the well-known result.

$$C_V = \frac{Nk_B\left[-T^2\left[\frac{\partial g(T)}{\partial T}\right]^2_{V,f} + g(T)T^2\frac{\partial^2 g(T)}{\partial T^2} + 2Tg(T)\frac{\partial g(T)}{\partial T}\right]}{\left[g(T)\right]^2}$$

$$C_V = nNK_B$$

Entropy: Eliminating f from equations $S = k_BfV\left[T\left[\frac{\partial g(T)}{\partial T}\right] + g(T)\right] - Nk_B\ln f$ and $N = \left[f\frac{\partial q(f,V,T)}{\partial f}\right]_{V,T} = fVg(T)$ we get,

$$S = \frac{k_BN}{g(T)}\left[T\left[\frac{\partial g(T)}{\partial T}\right] + g(T)\right] - Nk_B\ln\frac{N}{Vg(T)}$$

One can see that it is an extensive quantity.

Helmholtz free energy: Eliminating f from equations $F = Nk_BT\ln f - k_BT\ln\mathbb{Z}(f,V,T)$ $= Nk_BT\ln f - k_BTfVg(T)$ and $N = \left[f\frac{\partial q(f,V,T)}{\partial f}\right]_{V,T} = fVg(T)$ we get,

$$F = Nk_BT\left[\ln\frac{N}{Vg(T)} - 1\right]$$

Once again one can notice that it is an extensive quantity.

Derivation of Bose and Fermi Distribution Functions

Writing Grand canonical partition function in terms of single orbital partition functions offers a very interesting way to derive Bose and Fermi distributions. According to equation $\mathbb{Z} = \left\{\prod_r\sum_{N_r=0}^{\infty}\left(fe^{-\frac{E_r}{k_BT}}\right)^{N_r}\right\}$, single orbital partition function can be written as,

$$Z_1 = \sum_{N_r}\left(fe^{-\frac{Er}{kgT}}\right)^{N_r}$$

The summation in equation $\mu = -\frac{\alpha}{\beta}$ has two possibilities to choose N_r while taking the summation. In the case of Fermi-Dirac distribution N_r can take only two values $N_r = 0$ and 1, where as in the case of Bose Einstein Distribution N_r can take values from 0 to ∞. These two possible choices lead to two different results for the two distributions as discussed below:

Fermi-Dirac Distribution:

$$Z_1 = \left(fe^{-\frac{Er}{k_B T}} \right)^0 + \left(fe^{-\frac{Er}{k_B T}} \right)^1 = 1 + e^{\frac{\mu - E_r}{k_B T}}$$

Thus the grand partition function for the Fermi-Dirac Distribution becomes,

$$\mathbb{Z} = \left\{ \prod_r \left(1 + e^{\frac{\mu - E_r}{k_B T}} \right) \right\}$$

Therefore, q-potential can be obtained as,

$$q = \ln \mathbb{Z} = \sum_r \ln \left(1 + e^{\frac{\mu - E_r}{k_B T}} \right)$$

Bose-Einstein:

$$Z_1 = \sum_{N_r = 0}^{\infty} \left(fe^{-\frac{E_r}{k_B T}} \right)^{N_r} = 1 + e^{\frac{\mu - E_r}{k_B T}} + \left(e^{\frac{\mu - E_r}{k_B T}} \right)^2 + \cdots$$

This is a geometrical progression (G.P.) and is convergent provided $e^{\frac{\mu - E_r}{k_B T}} < 1$ for every E_r. This is possible only if $\mu < 0$, which is a very important result that for a bose system chemical potential is always negative. Thus above equation, which is a sum of a convergent series, can be written as

$$Z_1 = \frac{1}{\left(1 - e^{\frac{\mu - E_r}{k_B T}} \right)}$$

Thus the grand partition function for the Bose-Einstein Distribution becomes,

$$\mathbb{Z} = \left\{ \prod_r \left(1 - e^{\frac{\mu - E_r}{k_B T}} \right)^{-1} \right\}$$

Therefore, q-potential can be obtained as,

$$q = \ln \mathbb{Z} = -\sum_r \ln \left(1 - e^{\frac{\mu - E_r}{k_B T}} \right)$$

For Fermi-Dirac distribution and Bose-Einstein distribution together the q-potential can be written as,

$$q = \pm\sum_r \ln\left(1 \pm e^{\frac{\mu - E_r}{k_B T}}\right)$$

(+) sign gives Fermi-Dirac distribution and (-) gives Bose-Einstein distribution.

Since, $q(f,V,T) = \dfrac{PV}{k_B T}$

For the two distributions pressure can be obtained as,

$$P = \pm\frac{k_B T}{V}\sum_r \ln\left(1 \pm e^{\frac{\mu - E_r}{k_B T}}\right) = \pm\frac{k_B T}{V}\sum_r \ln\left(1 \pm f e^{\frac{-E_r}{k_B T}}\right)$$

Mean energy can be obtained from equation,

$$E = \langle E\rangle = -\left[\frac{\partial q(f,V,T)}{\partial \beta}\right]_{V,f} = k_B T^2\left[\frac{\partial q(f,V,T)}{\partial T}\right]_{V,f}$$

$$\langle E\rangle = -k_B T^2\left[\frac{\partial q(f,V,T)}{\partial T}\right]_{V,f} = \mp k_B T^2\left[\frac{\partial \sum_r \ln\left(1 \pm e^{\frac{\mu - E_r}{k_B T}}\right)}{\partial T}\right]_{V,f}$$

$$= \sum_r \frac{E_r}{\left(e^{\frac{E_r - \mu}{k_B T}} \pm 1\right)}$$

Average number of particles can be obtained from equation $F = E - TS \ldots -k_B T \ln\dfrac{\mathbb{Z}(f,V,T)}{f^N}$

$$\langle N\rangle = \left[f\frac{\partial q(f,V,T)}{\partial T}\right]_{V,T} = \sum_r \frac{1}{\left(f^{-1}e^{\frac{E_r}{k_B T}} \pm 1\right)} = \sum_r \frac{1}{\left(e^{\frac{E_r - \mu}{k_B T}} \pm 1\right)}$$

Mean occupation number of an orbital or level E_r

$$\langle nE_r\rangle = \left[f\frac{\partial q(f,V,T)}{\partial f}\right]_{V,T,\text{all other } E_r} = \frac{1}{\left(f^{-1}e^{\frac{E_r}{k_B T}} \pm 1\right)} = \frac{1}{\left(e^{\frac{E_r - \mu}{k_B T}} \pm 1\right)}$$

In the Maxwell Boltzmann case $e^{\frac{E_r-\mu}{k_BT}} >> 1$ for all values of single orbital energies including $E_r = 0$ which essential implies $e^{\frac{-\mu}{k_BT}} >> 1$ and μ must be negative and $|\mu| >> k_BT$. The Bose-Einstein Distribution and Fermi Dirac Distribution tend to so called Maxwell-Boltzmann,

$$\langle nE_r \rangle = e^{\frac{\mu-E_r}{k_BT}} \alpha\, e^{\frac{-E_r}{k_BT}}$$

Also mean occupation per orbital in Maxwell-Boltzmann reduces to the fact that $\langle nE_r \rangle << 1$, i.e. probability of occupation to be greater than 1 is negligible. Figure below shows occupation number per orbital state $\langle nE_r \rangle$ for the three cases. It is clear in the figure that for large value of $\frac{E_r-\mu}{k_BT}$ the Fermi-Dirac distribution curve (a) and Bose Einstein distribution curve (b) merges with the Maxwell-Boltzmann distribution curve (b). In the case of Fermi-Dirac distribution occupation number is never goes higher than 1. In the case of Bose-Einstein distribution as $E_r = \mu$, occupancy of that level becomes infinite leading to the phenomenon of Bose-Einstein Condensation.

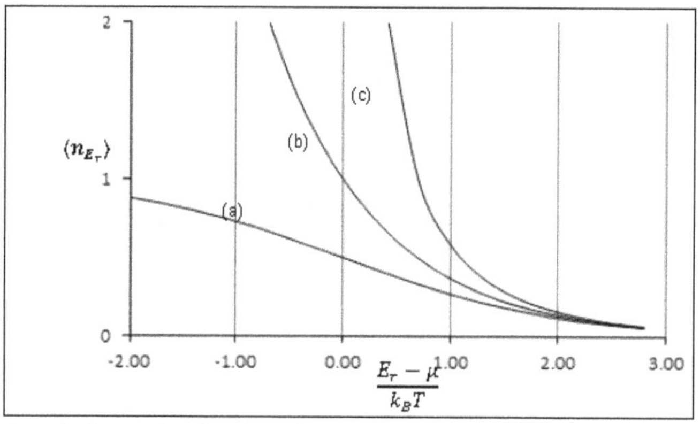

(a) Fermi Dirac Distribution (b) Maxwell Boltzmann Distribution (c) Bose Einstein Distribution for a non-interacting system.

Classical Mechanics Formulation of Statistical Mechanics

The ray approximation of classical mechanics breaks down when the de Broglie wavelength is not much smaller than other dimensions of the system. For non-relativistic particles, this wavelength is

$$\lambda = \frac{h}{p}$$

where h is Planck's constant and p is the momentum.

Again, this happens with electrons before it happens with heavier particles. For example, the electrons used by Clinton Davisson and Lester Germer in 1927, accelerated by 54 V, had a wavelength of 0.167 nm, which was long enough to exhibit a single diffraction side lobe when reflecting from the face of a nickel crystal with atomic spacing of 0.215 nm. With a larger vacuum chamber, it would seem relatively easy to increase the angular resolution from around a radian to a milliradian and see quantum diffraction from the periodic patterns of integrated circuit computer memory. More practical examples of the failure of classical mechanics on an engineering scale are conduction by quantum tunneling in tunnel diodes and very narrow transistor gates in integrated circuits.

Classical mechanics is the same extreme high frequency approximation as geometric optics. It is more often accurate because it describes particles and bodies with rest mass. These have more momentum and therefore shorter De Broglie wavelengths than massless particles, such as light, with the same kinetic energies.

Quantum Mechanical Formulation of Statistical Mechanics

Classically, state of the system is said to be specified once the position and momenta of all particles constituting the system gets specified. However, because of Heisenberg's Uncertainty principle, position and momenta cannot be specified simultaneously. In quantum mechanics, we deal with Hermitian operators which represent physically observable quantities with real eigen values. Some often encountered operators are $\hat{H}$, $\hat{L}$ and $\hat{S}$ representing respectively energy, angular momentum and spin angular momentum. Many physical quantities, like spin angular momentum, have no classical analogue. Each of the operators, operates on a vector space the so called Hilbert space. The state of the system at a certain time t is specified by a state vector $|\psi(t)\rangle$. Time evolution of the state $|\psi(t)\rangle$ is governed via the Schrodinger equation,

$$i\hbar\frac{\partial|\psi(t)\rangle}{\partial t}=\hat{H}|\psi(t)\rangle.$$

Knowing $|\psi(t)\rangle$ allows one to calculate all characteristics of the system equivalent to knowing the microstate of the system. The co-ordinate representation of the system is described by the wave function $\psi(q,t)=\langle q|\psi(t)\rangle$ an inner product of spatial and temporal function. Quantum mechanics is inherently probabilistic in nature such that $|\psi(q,t)|^2$ is the probability density. The probability that a system or a particle is located in a volume element dq around q is given by $|\psi(q,t)|^2\,dq$. Given a physical quantity, represented by an operator $\hat{O}$ say, As we operate $\hat{O}$ on a quantum state, one gets its eigen value $\langle\hat{O}\rangle$, which is the expectation value of $\hat{O}$, which is the average value of the operator obtained during the process of measurement given by,

$$\langle\hat{O}\rangle=\frac{\int\psi^*(q,t)\hat{O}\psi(q,t)d\tau}{\int\psi^*(q,t)\psi^*(q,t)d\tau}$$

Where $d\tau$ is the volume element of the co-ordinate space. If $\psi(q,t)$ is normalized than,

$$\int \psi^*(q,t)\psi(q,t)d\tau = \int |\psi(q,t)|^2 d\tau = 1.$$

For an operator $\hat{O}$ there can be a set of eigen values $\{o_n\}$ with a corresponding set of orthonormal eigen functions $\{\phi_n(q)\}$ forming a basis set, such that $\hat{O}\phi_n(q) = o_n\phi_n(q)$. The eigen function $\psi(q,t)$ can then be expanded in terms of this basis set as a result of superposition of basis eigen functions as,

$$\psi(q,t) = \sum_n a_n(t)\phi_n(q)$$

The physical significance of the coefficients lies in the fact that $|a_n(t)|^2$ gives the probability of getting in a measurement o_n corresponding to the eigenstate $\phi_n(q)$ out of the set of eigen function $\{\phi_n(q)\}$ where $a_n(t)$ is,

$$a_n(t) = \int \phi_n^*(q)\psi(q,t)d\tau$$

Density Matrix

We are now ready to apply these to a macroscopic system consisting of a large number of particles by using the notion of an ensemble. So we take an ensemble of N identical systems, where N >> 1. Since these systems are identical they have a common Hamiltonian $\hat{H}$. The states of the various systems making the ensemble is represented by the wave functions $\psi(q,t)$ where $q = \{q_i\}$ denote the position coordinates relevant to the system being considered. Therefore, $\psi(q,t)$ describes an equivalent of a phase point in phase space of a classical system.

Let us focus on the kth system of the ensemble. Let at a given time t the kth system of the ensemble happens to be in a normalized state $\psi^k(q,t)$, where k = 1, 2, ... N. The time evolution of this state function $\psi^k(q,t)$, is given by the time dependent Schrodinger equation,

$$i\hbar\frac{\partial\psi^k(t)}{\partial t} = \hat{H}\psi^k(t)$$

Where $\psi^k(t)$ can be expanded in terms of a set of orthonormal functions $\{\phi_n(q)\}$ as described in equation $\psi(q,t) = \sum_n a_n(t)\phi_n(q)$ in each system,

$$\psi^k(t) = \sum_n a_n^k(t)\phi_n(q)$$

Where,

$$a_n^k(t) = \int \phi_n^*(q)\psi^k(q,t)d\tau$$

Here $a\ (t)$ are the probability amplitudes of the system k to be in the state $\phi_n(q)$ and $|a_n^k(t)|^2$ is the probability amplitudes for the various systems of the ensemble in state $\phi_n(q)$. Sum of these probability amplitudes must be equal to 1.

$$\sum_n \left|a_n^k(t)\right|^2 = 1$$

Equation $i\hbar\frac{\partial\psi^k(t)}{\partial t} = \hat{H}\psi^k(t)$ can be written in terms of the co-efficient $a_n^k(t)$ so that time variation of these coefficients is given by,

$$\begin{aligned} i\hbar\frac{\partial a_n^k(t)}{\partial t} &= i\hbar\int \phi_n^*(q)\frac{\partial\psi^k(q,t)}{\partial t}d\tau \\ &= \int \phi_n^*(q)\hat{H}\psi^k(t)d\tau \\ &= \int \phi_n^*(q)\hat{H}\sum_n a_m^k(t)\phi_m(q)d\tau = \sum_m a_m^k(t)H_{mn} \end{aligned}$$

Where

$$H_{mn} = \int \phi_n^*(q)\hat{H}\phi_m(q)d\tau$$

Now let us define the density matrix by means of matrix elements,

$$\rho_{mn}(t) = \frac{1}{N}\sum_{k=1}^{N} a_m^k(t)a_n^{k^*}(t)$$

Since sum here is over all the systems in the ensemble divided by number of systems in the ensemble, it gives an ensemble average of the quantity $a_m^k(t)a_n^{k^*}(t)$. $m = n$ in above equation corresponds to the diagonal element $\rho_{nn}(t)$ which is the ensemble average of the probability density $|a_n(t)|^2$ which itself is a quantum average. So density matrix involves two averages, first one being a quantum mechanical average followed by an ensemble average.

So $\rho_{nn}(t)$ represents the probability that a system chosen at random from among the members of the ensemble shall be in the state φ_n at time t. Using equation $\sum_n \left|a_n^k(t)\right|^2 = 1$ we have,

$$\sum_n \rho_{nn}(t) = \frac{1}{N}\sum_{k=1}^{N}\sum_n a_n^k(t)a_n^{k^*}(t) = \frac{1}{N}\sum_{k=1}^{N}\sum_n \left|a_n^k(t)\right|^2 = 1$$

Let us now express the ensemble average of an operator $\hat{G}$

$$\langle G\rangle = \frac{1}{N}\sum_{k=1}^{N}\int \psi^{k*}\hat{G}\psi^k d\tau$$

This can further be written using equation $\psi^k(t) = \sum_n a_n^k(t)\phi_n(q)$ as,

$$\langle G\rangle = \frac{1}{N}\sum_{k=1}^{N}\left[\sum_{m.n} a_n^{*k}(t)a_m^k(t)G_{mn}\right]$$

Where,

$$G_{mn} = \int \phi_n^*(q)\hat{G}\phi_m(q)d\tau$$

Equation $\langle G\rangle = \frac{1}{N}\sum_{k=1}^{N}\left[\sum_{m.n} a_n^{*k}(t)a_m^k(t)G_{mn}\right]$ can be written as,

$$\langle G\rangle = \sum_{m.n}\left[\frac{1}{N}\sum_{k=1}^{N} a_n^{*k}(t)a_m^k(t)\right]G_{mn}$$

Or

$$\langle G\rangle = \sum_{m.n}\rho_{nm}G_{mn} = \sum_{m}\left(\hat{\rho}\hat{G}\right)_{mn} = \mathrm{Tr}\left(\hat{\rho}\hat{G}\right)$$

Note if $\hat{G} = \hat{1}$. Then, $\mathrm{Tr}\left(\hat{\rho}\right) = 1$

In general if ψ^k are not normalized,

$$\langle G\rangle = \frac{\mathrm{Tr}\left(\hat{\rho}\hat{G}\right)}{\mathrm{Tr}\left(\hat{\rho}\right)}$$

It is obvious from the above equation knowledge of density matrix allows us to calculate mean value of any physical property of the system. Density matrix is the quantum analogue of the density function of the classical system.

References

- Statistical-mechanics, science: britannica.com, Retrieved 12, February 2020
- Canonical-ensemble, science: britannica.com, Retrieved 30, May 2020

5

Relativity and Relativistic Mechanics

The branch of mechanics that is compatible with special relativity and general relativity is known as relativistic mechanics. It deals with the motion of bodies whose relative velocities are equal to the speed of light. This chapter discusses in detail the theories and methodologies related to relativistic mechanics such as theory of relativity, relativistic space-time, relativistic momentum, etc.

Relativistic mechanics, science concerned with the motion of bodies whose relative velocities approach the speed of light c, or whose kinetic energies are comparable with the product of their masses m and the square of the velocity of light, or mc^2. Such bodies are said to be relativistic, and when their motion is studied, it is necessary to take into account Einstein's special theory of relativity. As long as gravitational effects can be ignored, which is true so long as gravitational potential energydifferences are small compared with mc^2, the effects of Einstein's general theory of relativity may be safely ignored.

The bodies concerned may be sufficiently small that one may ignore their internal structure and size and regard them as point particles, in which case one speaks of relativistic point-particle mechanics; or one may need to take into account their internal structure, in which case one speaks of relativistic continuum mechanics. This part is concerned only with relativistic point-particle mechanics. It is also assumed that quantum mechanical effects are unimportant, otherwise relativistic quantum mechanics or relativistic quantum field theory—the latter theory being a quantum mechanical extension of relativistic continuum mechanics—would have to be considered. The condition that allows quantum effects to be safely ignored is that the sizes and separations of the bodies concerned are larger than their Compton wavelengths. (The Compton wavelength of a body of mass m is given by h/mc, where h is Planck's constant.) Despite these restrictions, there are nevertheless a number of situations in nature where relativistic mechanics is applicable. For example, it is essential to take into account the effects of relativity when calculating the motion of elementary particles accelerated to higher energies in particle accelerators, such as those at CERN (European Organization for Nuclear Research) near Geneva or at Fermilab (Fermi National Accelerator Laboratory) near Chicago. Moreover, such particles are caused to collide, thus creating further particles; although this creation process can only be understood through quantum mechanics, once the particles are well separated, they are subject to the laws of special relativity.

Similar remarks apply to cosmic rays that reach Earth from outer space. In some cases, these have energies as high as 10^{20} electron volts (eV). An electron of that energy has a velocity that differs from that of light by about 1 part in 10^{28}, as can be seen from the relativistic relation between energy and velocity, which will be given later. For a proton of the same energy, the velocity would differ from that of light by about 1 part in 10^{22}. At a more mundane level, relativistic mechanics must be used to calculate the energies of electrons or positrons emitted by the decay of radioactive nuclei. Astrophysicists need to use relativistic mechanics when dealing with the energy sources of stars, the energy released in supernova explosions, and the motion of electrons moving in the atmospheres of pulsars or when considering the hot big bang. At temperatures in the very early universe above 10^{10}kelvins (K), at which typical thermal energies kT (where k is Boltzmann's constant and T is temperature) are comparable with the rest mass energy of the electron, the primordial plasma must have been relativistic. Relativistic mechanics also must be considered when dealing with satellite navigational systems used, for example, by the military, such as the Global Positioning System (GPS). In this case, however, it is the purely kinematic effect on the rate of clocks on board the satellites (i.e., time dilation) that is important rather than the dynamic effects of relativity on the motion of the satellites themselves.

Relativistic Dynamics

Relativistic Energy and Momentum

There are a couple of (equivalent) ways to define momentum and energy in SR. One method uses conservation laws. If these laws are to remain valid in SR they must be true in every possible reference frame. However, if one does some simple thought experiments using the Newtonian definitions of momentum and energy, one sees that these quantities are not conserved in SR. One can rescue the idea of conservation by making some small modifications to the definitions to account for relativistic velocities. It is these new definitions which are taken as the correct ones for momentum and energy in SR.

The four-momentum of an object is straightforward, identical in form to the classical momentum, but replacing 3-vectors with 4-vectors:

$$\mathrm{P} = m_0 \mathrm{U} = (E / c, \mathrm{p})$$

The energy and momentum of an object with invariant mass m_0 (also called *rest mass*), moving with velocity v with respect to a given frame of reference, are respectively given by,

$$E = \gamma(\mathrm{v}) m_0 c^2$$
$$\mathrm{p} = \gamma(\mathrm{v}) m_0 \mathrm{v}$$

The factor of $\gamma(\mathrm{v})$ comes from the definition of the four-velocity. The appearance of the γ factor has an alternative way of being stated.

The kinetic energy, K, is defined as,

$$K = (\gamma - 1) m_0 c^2 = E - m_0 c^2,$$

And the speed as a function of kinetic energy is given by,

$$v = \frac{c\sqrt{K(K + 2m_0 c^2)}}{K + m_0 c^2} = \frac{c\sqrt{(E - m_0 c^2)(E + m_0 c^2)}}{E} = \frac{pc^2}{E}.$$

Rest Mass and Relativistic Mass

The quantity,

$$m = \gamma(\mathrm{v}) m_0$$

is often called the *relativistic mass* of the object in the given frame of reference.

This makes the relativistic relation between the spatial velocity and the spatial momentum look identical. However, this can be misleading, as it is *not* appropriate in special relativity in *all* circumstances. For instance, kinetic energy and force in special relativity can not be written exactly like their classical analogues by only replacing the mass with the relativistic mass. Moreover, under Lorentz transformations, this relativistic mass is not invariant, while the rest mass is. For this reason many people find it easier use the rest mass (thereby introduce γ through the 4-velocity or coordinate time), and discard the concept of relativistic mass.

Lev B. Okun suggested that "this terminology ... has no rational justification today", and should no longer be taught.

Other physicists, including Wolfgang Rindler and T. R. Sandin, have argued that relativistic mass is a useful concept and there is little reason to stop using it.

Some authors use m for relativistic mass and m_0 for rest mass, others simply use m for rest mass.

The energy and momentum of an object with invariant mass m_0 are related by the formulas,

$$E^2 - (pc)^2 = (m_0 c^2)^2$$
$$\mathrm{p}c^2 = E\mathrm{v}.$$

The first is referred to as the *relativistic energy–momentum relation*. It can be derived by considering that $\frac{v^2}{c^2}$ can be written as $\frac{p^2}{\gamma^2 m_0^2 c^2}$ where the denominator can be written as $\frac{E^2}{c^2}$. Now, gamma can be replaced in the expression of energy. While the energy E and the momentum p depend on the frame of reference in which they are measured, the quantity $E^2 - (pc)^2$ is invariant, and arises as $-c^2$ times the squared magnitude of the 4-momentum vector which is $-(m_0 c)^2$.

It should be noted that the invariant mass of a system,

$$m_{\mathrm{otot}} = \frac{\sqrt{E_{\mathrm{tot}}^2 - (p_{\mathrm{tot}} c)^2}}{c^2}$$

is different from the sum of the rest masses of the particles of which it is composed due to kinetic energy and binding energy. Rest mass is not a conserved quantity in special relativity unlike the situation in Newtonian physics. However, even if an object is changing internally, so long as it does not exchange energy with surroundings, then its rest mass will not change, and can be calculated with the same result in any frame of reference.

A particle whose rest mass is zero is called *massless*. Photons and gravitons are thought to be massless; and neutrinos are nearly so.

Mass–energy Equivalence

The relativistic energy–momentum equation holds for all particles, even for massless particles for which $m_0 = 0$. In this case:

$$E = pc$$

When substituted into $Ev = c^2p$, this gives $v = c$: massless particles (such as photons) always travel at the speed of light.

Notice that the rest mass of a composite system will generally be slightly different from the sum of the rest masses of its parts since, in its rest frame, their kinetic energy will increase its mass and their (negative) binding energy will decrease its mass. In particular, a hypothetical "box of light" would have rest mass even though made of particles which do not since their momenta would cancel.

Looking at the above formula for invariant mass of a system, one sees that, when a single massive object is at rest ($v = 0$, $p = 0$), there is a non-zero mass remaining: $m_0 = E/c^2$. The corresponding energy, which is also the total energy when a single particle is at rest, is referred to as "rest energy". In systems of particles which are seen from a moving inertial frame, total energy increases and so does momentum. However, for single particles the rest mass remains constant, and for systems of particles the invariant mass remain constant, because in both cases, the energy and momentum increases subtract from each other, and cancel. Thus, the invariant mass of systems of particles is a calculated constant for all observers, as is the rest mass of single particles.

The Mass of Systems and Conservation of Invariant Mass

For systems of particles, the energy–momentum equation requires summing the momentum vectors of the particles:

$$E^2 - \mathrm{p} \cdot \mathrm{p} c^2 = m_0^2 c^4$$

The inertial frame in which the momenta of all particles sums to zero is called the center of momentum frame. In this special frame, the relativistic energy–momentum equation has $p = 0$, and thus gives the invariant mass of the system as merely the total energy of all parts of the system, divided by c^2,

$$m_{0,\text{system}} = \sum_n E_n / c^2$$

This is the invariant mass of any system which is measured in a frame where it has zero total momentum, such as a bottle of hot gas on a scale. In such a system, the mass which the scale weighs is the invariant mass, and it depends on the total energy of the system. It is thus more than the sum of the rest masses of the molecules, but also includes all the totaled energies in the system as well. Like energy and momentum, the invariant mass of isolated systems cannot be changed so long as the system remains totally closed (no mass or energy allowed in or out), because the total relativistic energy of the system remains constant so long as nothing can enter or leave it.

An increase in the energy of such a system which is caused by translating the system to an inertial frame which is not the center of momentum frame, causes an increase in energy and momentum without an increase in invariant mass. $E = m_0c^2$, however, applies only to isolated systems in their center-of-momentum frame where momentum sums to zero.

Taking this formula at face value, we see that in relativity, mass is simply energy by another name (and measured in different units). In 1927 Einstein remarked about special relativity, "Under this theory mass is not an unalterable magnitude, but a magnitude dependent on (and, indeed, identical with) the amount of energy."

Closed (Isolated) Systems

In a "totally-closed" system (i.e., isolated system) the total energy, the total momentum, and hence the total invariant mass are conserved. Einstein's formula for change in mass translates to its simplest $\Delta E = \Delta mc^2$ form, however, only in non-closed systems in which energy is allowed to escape (for example, as heat and light), and thus invariant mass is reduced. Einstein's equation shows that such systems must lose mass, in accordance with the above formula, in proportion to the energy they lose to the surroundings. Conversely, if one can measure the differences in mass between a system before it undergoes a reaction which releases heat and light, and the system after the reaction when heat and light have escaped, one can estimate the amount of energy which escapes the system.

Chemical and Nuclear Reactions

In both nuclear and chemical reactions, such energy represents the difference in binding energies of electrons in atoms (for chemistry) or between nucleons in nuclei (in atomic reactions). In both cases, the mass difference between reactants and (cooled) products measures the mass of heat and light which will escape the reaction, and thus (using the equation) give the equivalent energy of heat and light which may be emitted if the reaction proceeds.

In chemistry, the mass differences associated with the emitted energy are around 10^{-9} of the molecular mass. However, in nuclear reactions the energies are so large that they are associated with mass differences, which can be estimated in advance, if the products and reactants have been weighed (atoms can be weighed indirectly by using atomic masses, which are always the same for each nuclide). Thus, Einstein's formula becomes important when one has measured the masses of different atomic nuclei. By looking at the difference in masses, one can predict which nuclei have stored energy that can be released by certain nuclear reactions, providing important information which was useful in the development of nuclear energy and, consequently, the nuclear bomb. Historically, for example, Lise Meitner was able to use the mass differences in nuclei to estimate that there was enough energy available to make nuclear fission a favorable process. The implications of this special form of Einstein's formula have thus made it one of the most famous equations in all of science.

Center of Momentum Frame

The equation $E = m_0c^2$ applies only to isolated systems in their center of momentum frame. It has been popularly misunderstood to mean that mass may be *converted* to energy, after which the *mass* disappears. However, popular explanations of the equation as applied to systems include open (non-isolated) systems for which heat and light are allowed to escape, when they otherwise would have contributed to the mass (invariant mass) of the system.

Historically, confusion about mass being "converted" to energy has been aided by confusion between mass and "matter", where matter is defined as fermion particles. In such a definition, electromagnetic radiation and kinetic energy (or heat) are not considered "matter". In some situations, matter may indeed be converted to non-matter forms of energy, but in all these situations, the matter and non-matter forms of energy still retain their original mass.

For isolated systems (closed to all mass and energy exchange), mass never disappears in the center of momentum frame, because energy cannot disappear. Instead, this equation, in context, means only that when any energy is added to, or escapes from, a system in the center-of-momentum frame, the system will be measured as having gained or lost mass, in proportion to energy added or removed. Thus, in theory, if an atomic bomb were placed in a box strong enough to hold its blast, and detonated upon a scale, the mass of this closed system would not change, and the scale would not move. Only when a transparent "window" was opened in the super-strong plasma-filled box, and light and heat were allowed to escape in a beam, and the bomb components to cool, would the system lose the mass associated with the energy of the blast. In a 21 kiloton bomb, for example, about a gram of light and heat is created. If this heat and light were allowed to escape, the remains of the bomb would lose a gram of mass, as it cooled. In this thought-experiment, the light and heat carry away the gram of mass, and would therefore deposit this gram of mass in the objects that absorb them.

Angular Momentum

In relativistic mechanics, the time-varying mass moment,

$$\mathrm{N} = m(\mathrm{x} - t\mathrm{v})$$

and orbital 3-angular momentum,

$$\mathrm{L} = \mathrm{x} \times \mathrm{p}$$

of a point-like particle are combined into a four-dimensional bivector in terms of the 4-position X and the 4-momentum P of the particle:

$$\mathrm{M} = \mathrm{X} \wedge \mathrm{P}$$

where, $\wedge$ denotes the exterior product. This tensor is additive: the total angular momentum of a system is the sum of the angular momentum tensors for each constituent of the system. So, for an assembly of discrete particles one sums the angular momentum tensors over the particles, or integrates the density of angular momentum over the extent of a continuous mass distribution.

Each of the six components forms a conserved quantity when aggregated with the corresponding components for other objects and fields.

Force

In special relativity, Newton's second law does not hold in the form F = ma, but it does if it is expressed as,

$$F = \frac{dp}{dt}$$

where p = $\gamma(v) m_o v$ is the momentum as defined above and m_o is the invariant mass. Thus, the force is given by,

$$F = \gamma(v)^3 m_o a_{\parallel} + \gamma(v) m_o a_{\perp}$$

Derivation

Starting from,

$$F = m_o \frac{d(\gamma(v)v)}{dt} = m_o\left(\frac{d\gamma(v)}{dt}v + \gamma(v)\frac{dv}{dt}\right).$$

Carrying out the derivatives gives,

$$F = \frac{\gamma(v)^3 m_o}{c^2}(v\cdot a)v + \gamma(v) m_o a.$$

If the acceleration is separated into the part parallel to the velocity ($a_{\parallel}$) and the part perpendicular to it ($a_{\perp}$), so that:

$$a = a_{\parallel} + a_{\perp}, \quad v\cdot a_{\perp} = 0, \quad v\cdot a = v\cdot a_{\parallel},$$

one gets,

$$F = \frac{\gamma(v)^3 m_o}{c^2}(v\cdot a_{\parallel})v + \gamma(v) m_o (a_{\perp} + a_{\parallel}).$$

By construction $a_{\parallel}$ and v are parallel, so $(v\cdot a_{\parallel})v$ is a vector with magnitude $v^2 a_{\parallel}$ in the direction of v (and hence $a_{\parallel}$) which allows the replacement:

$$(v\cdot a_{\parallel})v = v^2 a_{\parallel}$$

then,

$$\begin{aligned}F &= \frac{\gamma(v)^3 m_o v^2}{c^2} a_{\parallel} + \gamma(v) m_o (a_{\parallel} + a_{\perp}) \\ &= \gamma(v)^3 m_o\left(\frac{v^2}{c^2} + \frac{1}{\gamma(v)^2}\right) a_{\parallel} + \gamma(v) m_o a_{\perp} \\ &= \gamma(v)^3 m_o\left(\frac{v^2}{c^2} + 1 - \frac{v^2}{c^2}\right) a_{\parallel} + \gamma(v) m_o a_{\perp} \\ &= \gamma(v)^3 m_o a_{\parallel} + \gamma(v) m_o a_{\perp}\end{aligned}$$

Consequently, in some old texts, $\gamma(v)^3 m_o$ is referred to as the *longitudinal mass*, and $\gamma(v) m_o$ is referred to as the *transverse mass*, which is numerically the same as the relativistic mass.

If one inverts this to calculate acceleration from force, one gets,

$$a = \frac{1}{m_o \gamma(v)} \left(F - \frac{(v \cdot F)v}{c^2} \right).$$

The force described in this section is the classical 3-D force which is not a four-vector. This 3-D force is the appropriate concept of force since it is the force which obeys Newton's third law of motion. It should not be confused with the so-called four-force which is merely the 3-D force in the comoving frame of the object transformed as if it were a four-vector. However, the density of 3-D force (linear momentum transferred per unit four-volume) *is* a four-vector (density of weight +1) when combined with the negative of the density of power transferred.

Torque

The torque acting on a point-like particle is defined as the derivative of the angular momentum tensor given above with respect to proper time:

$$\Gamma = \frac{dM}{d\tau} = X \wedge F$$

or in tensor components:

$$\Gamma_{\alpha\beta} = X_\alpha F_\beta - X_\beta F_\alpha$$

where F is the 4d force acting on the particle at the event X. As with angular momentum, torque is additive, so for an extended object one sums or integrates over the distribution of mass.

Kinetic Energy

The *work-energy theorem* says the change in kinetic energy is equal to the work done on the body. In special relativity:

$$\Delta K = W = [\gamma_1 - \gamma_0] m_0 c^2.$$

Derivation

$$\ddot{A}K = W = \int_{x_0}^{x_1} F \cdot dx$$

$$= \int_{t_0}^{t_1} \frac{d}{dt}(\gamma m_0 v) \cdot v dt$$

$$= \gamma m_0 v \cdot v \Big|_{t_0}^{t_1} - \int_{t_0}^{t_1} \gamma m_0 v \cdot \frac{dv}{dt} dt$$

$$= \gamma m_0 v^2 \Big|_{t_0}^{t_1} - m_0 \int_{v_0}^{v_1} \gamma v dv$$

$$= m_0 \left(\gamma v^2 \Big|_{t_0}^{t_1} - c^2 \int_{v_0}^{v_1} \frac{2v / c^2}{2\sqrt{1 - v^2 / c^2}} dv \right)$$

$$= m_0 \left(\frac{v^2}{\sqrt{1 - v^2 / c^2}} + c^2 \sqrt{1 - v^2 / c^2} \right) \Bigg|_{t_0}^{t_1}$$

$$= \frac{m_0 c^2}{\sqrt{1 - v^2 / c^2}} \Bigg|_{t_0}^{t_1}$$

$$= \gamma m_0 c^2 \Big|_{t_0}^{t_1}$$

$$= \gamma_1 m_0 c^2 - \gamma_0 m c^2.$$

If in the initial state the body was at rest, so $v_0 = 0$ and $\gamma_0(v_0) = 1$, and in the final state it has speed $v_1 = v$, setting $\gamma_1(v_1) = \gamma(v)$, the kinetic energy is then;

$$K = [\gamma(v) - 1] m_0 c^2,$$

a result that can be directly obtained by subtracting the rest energy m_0c^2 from the total relativistic energy $\gamma(v)m_0c^2$.

Newtonian Limit

The Lorentz factor $\gamma(v)$ can be expanded into a Taylor series or binomial series for $(v/c)^2 < 1$, obtaining:

$$\gamma = \frac{1}{\sqrt{1 - (v/c)^2}} = \sum_{n=0}^{\infty} \left(\frac{v}{c}\right)^{2n} \prod_{k=1}^{n} \left(\frac{2k-1}{2k}\right) = 1 + \frac{1}{2}\left(\frac{v}{c}\right)^2 + \frac{3}{8}\left(\frac{v}{c}\right)^4 + \frac{5}{16}\left(\frac{v}{c}\right)^6 + \cdots$$

and consequently,

$$E - m_0 c^2 = \frac{1}{2} m_0 v^2 + \frac{3}{8}\frac{m_0 v^4}{c^2} + \frac{5}{16}\frac{m_0 v^6}{c^4} + \cdots;$$

$$\mathrm{p} = m_0 \mathrm{v} + \frac{1}{2}\frac{m_0 v^2 \mathrm{v}}{c^2} + \frac{3}{8}\frac{m_0 v^4 \mathrm{v}}{c^4} + \frac{5}{16}\frac{m_0 v^6 \mathrm{v}}{c^6} + \cdots.$$

For velocities much smaller than that of light, one can neglect the terms with c^2 and higher in the denominator. These formulas then reduce to the standard definitions of Newtonian kinetic energy and momentum. This is as it should be, for special relativity must agree with Newtonian mechanics at low velocities.

Theory of Relativity

The Theory of Relativity, proposed by the Jewish physicist Albert Einstein (1879-1955) in the early part of the 20th century, is one of the most significant scientific advances of our time. Although the concept of relativity was not introduced by Einstein, his major contribution was the recognition that the speed of light in a vacuum is constant and an absolute physical boundary for motion. This does not have a major impact on a person's day-to-day life since we travel at speeds much slower than light speed. For objects travelling near light speed, however, the theory of relativity states that objects will move slower and shorten in length from the point of view of an observer on Earth. Einstein also derived the famous equation, $E = mc^2$, which reveals the equivalence of mass and energy.

When Einstein applied his theory to gravitational fields, he derived the "curved space-time continuum" which depicts the dimensions of space and time as a two-dimensional surface where massive objects create valleys and dips in the surface. This aspect of relativity explained the phenomena of light bending around the sun, predicted black holes as well as the Cosmic Microwave Background Radiation (CMB) -- a discovery rendering fundamental anomalies in the classic Steady-State hypothesis. For his work on relativity, the photoelectric effect and blackbody radiation, Einstein received the Nobel Prize in 1921.

Theory of Relativity – The Basics

Physicists usually dichotomize the Theory of Relativity into two parts.

- The first is the Special Theory of Relativity, which essentially deals with the question of whether rest and motion are relative or absolute, and with the consequences of Einstein's conjecture that they are relative.
- The second is the General Theory of Relativity, which primarily applies to particles as they accelerate, particularly due to gravitation, and acts as a radical revision of Newton's theory, predicting important new results for fast-moving and/or very massive bodies. The General Theory of Relativity correctly reproduces all validated predictions of Newton's theory, but expands on our understanding of some of the key principles. Newtonian physics had previously hypothesised that gravity operated through empty space, but the theory lacked explanatory power as far as how the distance and mass of a given object could be transmitted through space. General relativity irons out this paradox, for it shows that objects continue to move in a straight line in space-time, but we observe the motion as acceleration because of the curved nature of space-time.

Einstein's theories of both special and general relativity have been confirmed to be accurate to a very high degree over recent years, and the data has been shown to corroborate many key predictions; the most famous being the solar eclipse of 1919 bearing testimony that the light of stars is indeed deflected by the sun as the light passes near the sun on its way to earth. The total solar eclipse allowed astronomers to -- for the first time -- analyse starlight near the edge of the sun, which had been previously inaccessible to observers due to the intense brightness of the sun. It also predicted the rate at which two neutron stars orbiting one another will move toward each other. When this phenomenon was first documented, general relativity proved itself accurate to better than a trillionth of a percent precision, thus making it one of the best confirmed principles in all of physics.

Applying the principle of general relativity to our cosmos reveals that it is not static. Edwin Hubble (1889-1953) demonstrated in 1928 that the Universe is expanding, showing beyond reasonable doubt that the Universe sprang into being a finite time ago. The most common contemporary interpretation of this expansion is that this began to exist from the moment of the Big Bang some 13.7 billion years ago. However this is not the only plausible cosmological model which exists in academia, and many creation physicists such as Russell Humphreys and John Hartnett have devised models operating with a biblical framework, which -- to date -- have withstood the test of criticism from the most vehement of opponents.

A Testament to Creation

Using the observed cosmic expansion conjunctively with the general theory of relativity, we can infer from the data that the further back into time one looks, the universe ought to diminish in size accordingly. However, this cannot be extrapolated indefinitely. The universe's expansion helps us to appreciate the direction in which time flows. This is referred to as the Cosmological arrow of time, and implies that the future is -- by definition -- the direction towards which the universe increases in size. The expansion of the universe also gives rise to the second law of thermodynamics, which states that the overall entropy (or disorder) in the Universe can only increase with time because the amount of energy available for work deteriorates with time. If the universe was eternal, therefore, the amount of usable energy available for work would have already been exhausted. Hence it follows that at one point the entropy value was at absolute 0 (most ordered state at the moment of creation) and the entropy has been increasing ever since -- that is, the universe at one point was fully "wound up" and has been winding down ever since. This has profound theological implications, for it shows that time itself is necessarily finite. If the universe were eternal, the thermal energy in the universe would have been evenly distributed throughout the cosmos, leaving each region of the cosmos at uniform temperature (at very close to absolute 0), rendering no further work possible.

The General Theory of Relativity demonstrates that time is linked, or related, to matter and space, and thus the dimensions of time, space, and matter constitute what we would call a continuum. They must come into being at precisely the same instant. Time itself cannot exist in the absence of matter and space. From this, we can infer that the uncaused first cause must exist outside of the four dimensions of space and time, and possess eternal, personal, and intelligent qualities in order to possess the capabilities of intentionally space, matter -- and indeed even time itself -- into being.

Moreover, the very physical nature of time and space also suggest a Creator, for infinity and eternity must necessarily exist from a logical perspective. The existence of time implies eternity (as time has a beginning and an end), and the existence of space implies infinity. The very concepts of infinity and eternity infer a Creator because they find their very state of being in God, who transcends both and simply is.

Special Relativity

Special relativity is a fundamental physics theory about space and time that was developed by Albert Einstein in 1905 as a modification of Newtonian physics. It was created to deal with some pressing theoretical and experimental issues in the physics of the time involving light and electrodynamics. The predictions of special relativity correspond closely to those of Newtonian physics at speeds which are low in comparison to that of light, but diverge rapidly for speeds which are a significant fraction of the speed of light. Special relativity has been experimentally tested on numerous occasions since its inception, and its predictions have been verified by those tests.

Albert Einstein during a lecture in Vienna.

Einstein postulated that the speed of light is the same for all observers, irrespective of their motion relative to the light source. This was in total contradiction to classical mechanics, which had been accepted for centuries. Einstein's approach was based on thought experiments and calculations. In 1908, Hermann Minkowski reformulated the theory based on different postulates of a more geometrical nature. His approach depended on the existence of certain interrelations between space and time, which were considered completely separate in classical physics. This reformulation set the stage for further developments of physics.

Special relativity makes numerous predictions that are incompatible with Newtonian physics (and everyday intuition). The first such prediction described by Einstein is called the relativity of simultaneity, under which observers who are in motion with respect to each other may disagree on whether two events occurred at the same time or one occurred before the other. The other major predictions of special relativity are time dilation (under which a moving clock ticks more slowly than when it is at rest with respect the observer), length contraction (under which a moving rod may be found to be shorter than when it is at rest with respect to the observer), and the equivalence of mass and energy (written as $E=mc^2$). Special relativity predicts a non-linear velocity addition formula, which prevents speeds greater than that of light from being observed. Special relativity also explains why Maxwell's equations of electromagnetism are correct in any frame of reference, and how an electric field and a magnetic field are two aspects of the same thing.

Special relativity has received experimental support in many ways, and it has been proven far more accurate than Newtonian mechanics. The most famous experimental support is the Michelson-Morley experiment, the results of which (showing that the speed of light is a constant) was one factor that motivated the formulation of the theory of special relativity. Other significant tests are the Fizeau experiment (which was first done decades before special relativity was proposed), the detection of the transverse Doppler effect, and the Haefele-Keating experiment. Today, scientists are so comfortable with the idea that the speed of light is always the same that the meter is now defined as being the distance traveled by light in $1/299,792,458^{th}$ of a second. This means that the speed of light is now defined as being $299,792,458$ m/s.

Reference Frames and Galilean Relativity: A Classical Prelude

A reference frame is simply a selection of what constitutes stationary objects. Once the velocity of a certain object is arbitrarily defined to be zero, the velocity of everything else in the universe can be measured relative to it. When a train is moving at a constant velocity past a platform, one may either say that the platform is at rest and the train is moving or that the train is at rest and the platform is moving past it. These two descriptions correspond to two different reference frames. They are respectively called the rest frame of the platform and the rest frame of the train (sometimes simply the platform frame and the train frame).

The question naturally arises, can different reference frames be physically differentiated? In other words, can one conduct some experiments to claim that "we are now in an absolutely stationary reference frame?" Aristotle thought that all objects tend to cease moving and become at rest if there were no forces acting on them. Galileo challenged this idea and argued that the concept of absolute motion was unreal. All motion was relative. An observer who couldn't refer to some isolated object (if, say, he was imprisoned inside a closed spaceship) could never distinguish whether according to some external observer he was at rest or moving with constant velocity. Any experiment he could conduct would give the same result in both cases. However, accelerated reference frames are experimentally distinguishable. For example, if an astronaut moving in free space saw that the tea in his tea-cup was slanted rather than horizontal, he would be able to infer that his spaceship was accelerated. Thus not all reference frames are equivalent, but people have a class of reference frames, all moving at uniform velocity with respect to each other, in all of which Newton's first law holds. These are called the inertial reference frames and are fundamental to both classical mechanics and SR. Galilean relativity thus states that the laws of physics can not depend on absolute velocity, they must stay the same in any inertial reference frame. Galilean relativity is thus a fundamental principle in classical physics.

Mathematically, it says that if one transforms all velocities to a different reference frame, the laws of physics must be unchanged. What is this transformation that must be applied to the velocities? Galileo gave the common-sense "formula" for adding velocities: If

1. Particle P is moving at velocity v with respect to reference frame A,
2. Reference frame A is moving at velocity u with respect to reference frame B, then
3. The velocity of P with respect to B is given by v + u.

The formula for transforming coordinates between different reference frames is called the Galilean transformation. The principle of Galilean relativity then demands that laws of physics be unchanged if the Galilean transformation is applied to them. Laws of classical mechanics, like Newton's second law, obey this principle because they have the same form after applying the transformation. As Newton's law involves the derivative of velocity, any constant velocity added in a Galilean transformation to a different reference frame contributes nothing (the derivative of a constant is zero). Addition of a time-varying velocity (corresponding to an accelerated reference frame) will however change the formula, since Galilean relativity only applies to non-accelerated inertial reference frames.

Time is the same in all reference frames because it is absolute in classical mechanics. All observers measure exactly the same intervals of time and there is such a thing as an absolutely correct clock.

Invariance of Length: The Euclidean Picture

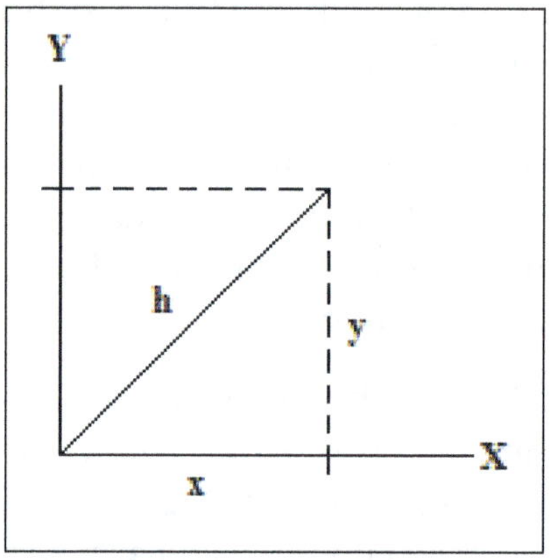

Pythagoras theorem.

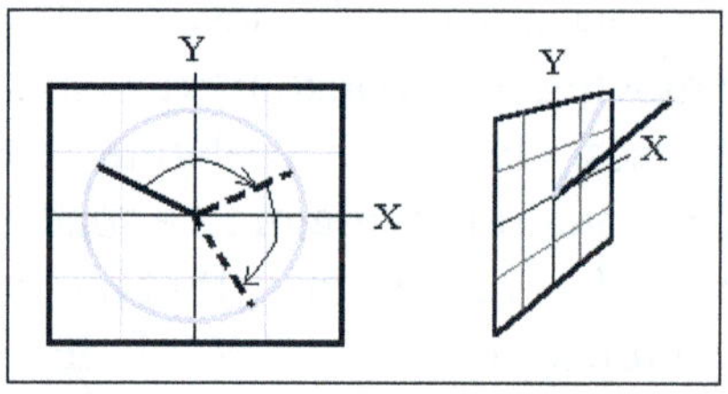

The length of an object is constant on the plane during rotations on the plane but not during rotations out of the plane.

In special relativity, space and time are joined into a unified four-dimensional continuum called spacetime. To gain a sense of what spacetime is like, we must first look at the Euclidean space of Newtonian physics.

This approach to the theory of special relativity begins with the concept of "length." In everyday experience, it seems that the length of objects remains the same no matter how they are rotated or moved from place to place; as a result the simple length of an object doesn't appear to change or is "invariant." However, as is shown in the illustrations below, what is actually being suggested is that length seems to be invariant in a three-dimensional coordinate system.

The length of a line in a two-dimensional Cartesian coordinate system is given by Pythagoras' theorem:

$$h^2 = x^2 + y^2.$$

One of the basic theorems of vector algebra is that the length of a vector does not change when it is rotated. However, a closer inspection tells us that this is only true if we consider rotations confined to the plane. If we introduce rotation in the third dimension, then we can tilt the line out of the plane. In this case the projection of the line on the plane will get shorter. Does this mean length is not invariant? Obviously not. The world is three-dimensional and in a 3D Cartesian coordinate system the length is given by the three-dimensional version of Pythagoras's theorem:

$$k^2 = x^2 + y^2 + z^2.$$

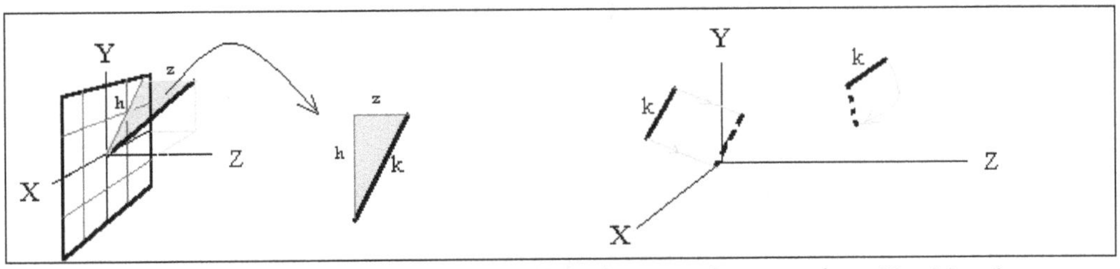

Invariance in a 3D coordinate system: Pythagoras theorem gives $k^2 = h^2 + z^2$ but $h^2 = x^2 + y^2$ therefore $k^2 = x^2 + y^2 + z^2$. The length of an object is constant whether it is rotated or moved from one place to another in a 3D coordinate system.

This is invariant under all rotations. The apparent violation of invariance of length only happened because we were "missing" a dimension. It seems that, provided all the directions in which an object can be tilted or arranged are represented within a coordinate system, the length of an object does not change under rotations. A 3-dimensional coordinate system is enough in classical mechanics because time is assumed absolute and independent of space in that context. It can be considered separately.

Note that invariance of length is not ordinarily considered a dynamic principle, not even a theorem. It is simply a statement about the fundamental nature of space itself. Space as we ordinarily conceive it is called a three-dimensional Euclidean space, because its geometrical structure is described by the principles of Euclidean geometry. The formula for distance between two points is a fundamental property of an Euclidean space, it is called the Euclidean metric tensor (or simply the Euclidean metric). In general, distance formulas are called metric tensors.

Note that rotations are fundamentally related to the concept of length. In fact, one may define length or distance to be that which stays the same (is invariant) under rotations, or define rotations to be that which keep the length invariant. Given any one, it is possible to find the other. If we know the distance formula, we can find out the formula for transforming coordinates in a rotation. If, on the other hand, we have the formula for rotations then we can find out the distance formula.

Postulates of Special Relativity

Einstein developed Special Relativity on the basis of two postulates:

- First postulate—Special principle of relativity—The laws of physics are the same in all inertial frames of reference. In other words, there are no privileged inertial frames of reference.
- Second postulate—Invariance of *c*—The speed of light in a vacuum is independent of the motion of the light source.

Special Relativity can be derived from these postulates, as was done by Einstein in 1905. Einstein's postulates are still applicable in the modern theory but the origin of the postulates is more explicit. It was shown above how the existence of a universally constant velocity (the speed of light) is a consequence of modeling the universe as a particular four dimensional space having certain specific properties. The principle of relativity is a result of Minkowski structure being preserved under Lorentz transformations, which are *postulated* to be the physical transformations of inertial reference frames.

Minkowski Formulation: Introduction of Spacetime

Hermann Minkowski.

After Einstein derived special relativity formally from the counterintuitive proposition that the speed of light is the same to all observers, the need was felt for a more satisfactory formulation. Minkowski, building on mathematical approaches used in non-Euclidean geometry and the mathematical work of Lorentz and Poincaré, realized that a geometric approach was the key. Minkowski showed in 1908 that Einstein's new theory could be explained in a natural way if the concept of separate space and time is replaced with one four-dimensional continuum called spacetime. This was a groundbreaking concept, and Roger Penrose has said that relativity was not truly complete until Minkowski reformulated Einstein's work.

The concept of a four-dimensional space is hard to visualize. It may help at the beginning to think simply in terms of coordinates. In three-dimensional space, one needs three real numbers to refer to a point. In the Minkowski space, one needs four real numbers (three space coordinates and one time coordinate) to refer to a point at a particular instant of time. This point at a particular instant of time, specified by the four coordinates, is called an event. The distance between two different events is called the spacetime interval.

A path through the four-dimensional spacetime, usually called Minkowski space, is called a world line. Since it specifies both position and time, a particle having a known world line has a completely determined trajectory and velocity. This is just like graphing the displacement of a particle moving in a straight line against the time elapsed. The curve contains the complete motional information of the particle.

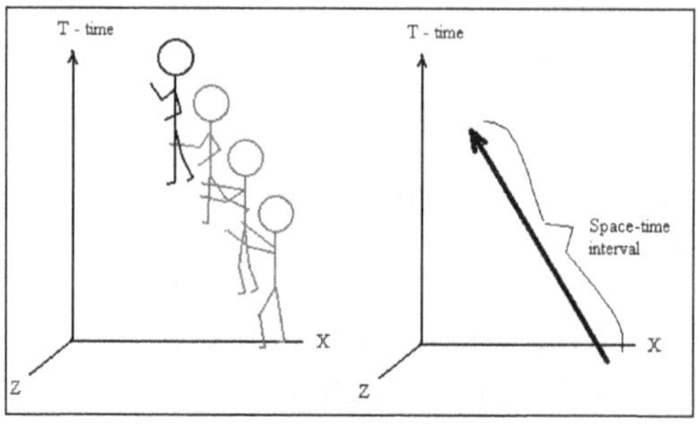

The spacetime interval.

In the same way as the measurement of distance in 3D space needed all three coordinates we must include time as well as the three space coordinates when calculating the distance in Minkowski space (henceforth called M). In a sense, the spacetime interval provides a combined estimate of how far two events occur in space as well as the time that elapses between their occurrence.

But there is a problem. Time is related to the space coordinates, but they are not equivalent. Pythagoras's theorem treats all coordinates on an equal footing. We can exchange two space coordinates without changing the length, but we can not simply exchange a space coordinate with time, they are fundamentally different. It is an entirely different thing for two events to be separated in space and to be separated in time. Minkowski proposed that the formula for distance needed a change. He found that the correct formula was actually quite simple, differing only by a sign from Pythagoras's theorem:

$$s^2 = x^2 + y^2 + z^2 - (ct)^2$$

where c is a constant and t is the time coordinate. Multiplication by c, which has the dimension ms^{-1}, converts the time to units of length and this constant has the same value as the speed of light. So the spacetime interval between two distinct events is given by,

$$s^2 = (x_2 - x_1)^2 + (y_2 - y_1)^2 + (z_2 - z_1)^2 - c^2(t_2 - t_1)^2.$$

There are two major points to be noted. Firstly, time is being measured in the same units as length by multiplying it by a constant conversion factor. Secondly, and more importantly, the time-coordinate has a different sign than the space coordinates. This means that in the four-dimensional spacetime, one coordinate is different from the others and influences the distance differently. This new 'distance' may be zero or even negative. This new distance formula, called the metric of the spacetime, is at the heart of relativity. This distance formula is called the metric tensor of M. This minus sign means that a lot of our intuition about distances can not be directly carried over into spacetime intervals. For example, the spacetime interval between two events separated both in time and space may be zero. From now on, the terms distance formula and metric tensor will be used interchangeably, as will be the terms Minkowski metric and spacetime interval.

In Minkowski spacetime the spacetime interval is the invariant length, the ordinary 3D length is not required to be invariant. The spacetime interval must stay the same under rotations, but ordinary lengths can change. Just like before, we were missing a dimension. Note that everything this far are merely definitions. We define a four-dimensional mathematical construct which has a special formula for distance, where distance means that which stays the same under rotations (alternatively, one may define a rotation to be that which keeps the distance unchanged).

Now comes the physical part. Rotations in Minkowski space have a different interpretation than ordinary rotations. These rotations correspond to transformations of reference frames. Passing from one reference frame to another corresponds to rotating the Minkowski space. An intuitive justification for this is given below, but mathematically this is a dynamical postulate just like assuming that physical laws must stay the same under Galilean transformations (which seems so intuitive that we don't usually recognize it to be a postulate).

Since by definition rotations must keep the distance same, passing to a different reference frame must keep the spacetime interval between two events unchanged. This requirement can be used to derive an explicit mathematical form for the transformation that must be applied to the laws of physics (compare with the application of Galilean transformations to classical laws) when shifting reference frames. These transformations are called the Lorentz transformations. Just like the Galilean transformations are the mathematical statement of the principle of Galilean relativity in classical mechanics, the Lorentz transformations are the mathematical form of Einstein's principle of relativity. Laws of physics must stay the same under Lorentz transformations. Maxwell's equations and Dirac's equation satisfy this property, and hence, they are relativistically correct laws (but classically incorrect, since they don't transform correctly under Galilean transformations).

With the statement of the Minkowski metric, the common name for the distance formula given above, the theoretical foundation of special relativity is complete. The entire basis for special relativity can be summed up by the geometric statement "changes of reference frame correspond to rotations in the 4D Minkowski spacetime, which is defined to have the distance formula given above." The unique dynamical predictions of SR stem from this geometrical property of spacetime. Special relativity may be said to be the physics of Minkowski spacetime. In this case of spacetime, there are six independent rotations to be considered. Three of them are the standard rotations on a plane in two directions of space. The other three are rotations in a plane of both space and time: These rotations correspond to a change of velocity, and are described by the traditional Lorentz transformations.

One can replace distance formulas with rotation formulas. Instead of starting with the invariance of the Minkowski metric as the fundamental property of spacetime, one may state (as was done in classical physics with Galilean relativity) the mathematical form of the Lorentz transformations and require that physical laws be invariant under these transformations. This makes no reference to the geometry of spacetime, but will produce the same result. This was in fact the traditional approach to SR, used originally by Einstein himself. However, this approach is often considered to offer less insight and be more cumbersome than the more natural Minkowski formalism.

Reference Frames and Lorentz Transformations: Relativity Revisited

We have already discussed that in classical mechanics coordinate frame changes correspond to Galilean transfomations of the coordinates. Is this adequate in the relativistic Minkowski picture?

Suppose there are two people, Bill and John, on separate planets that are moving away from each other. Bill and John are on separate planets so they both think that they are stationary. John draws a graph of Bill's motion through space and time and this is shown in the illustration below.

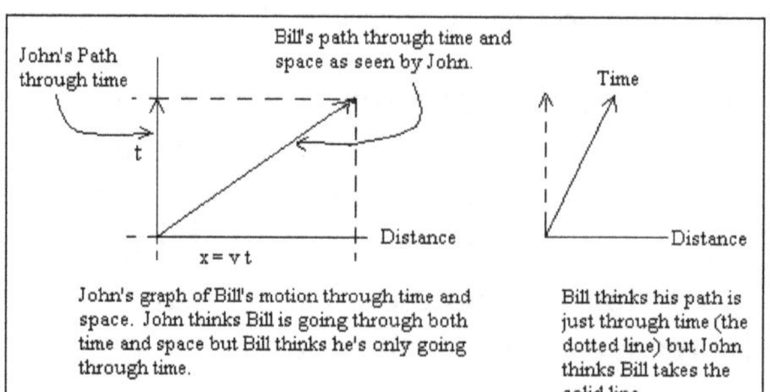

John's view of Bill and Bill's view of himself.

John sees that Bill is moving through space as well as time but Bill thinks he is moving through time alone. Bill would draw the same conclusion about John's motion. In fact, these two views, which would be classically considered a difference in reference frames, are related simply by a co-ordinate transformation in M. Bill's view of his own world line and John's view of Bill's world line are related to each other simply by a rotation of coordinates. One can be transformed into the other by a rotation of the time axis. Minkowski geometry handles transformations of reference frames in a very natural way.

Changes in reference frame, represented by velocity transformations in classical mechanics, are represented by rotations in Minkowski space. These rotations are called Lorentz transformations. They are different from the Galilean transformations because of the unique form of the Minkowski metric. The Lorentz transformations are the relativistic equivalent of Galilean transformations. Laws of physics, in order to be relativistically correct, must stay the same under Lorentz transformations. The physical statement that they must be same in all inertial reference frames remains unchanged, but the mathematical transformation between different reference frames changes. Newton's laws of motion are invariant under Galilean rather than Lorentz transformations, so they are immediately recognizable as non-relativistic laws and must be discarded in relativistic physics. Schrödinger's equation is also non-relativistic.

Maxwell's equations are trickier. They are written using vectors and at first glance appear to transform correctly under Galilean transformations. But on closer inspection, several questions are apparent that can not be satisfactorily resolved within classical mechanics. They are indeed invariant under Lorentz transformations and are relativistic, even though they were formulated before the discovery of special relativity. Classical electrodynamics can be said to be the first relativistic theory in physics. To make the relativistic character of equations apparent, they are written using 4-component vector like quantities called 4-vectors. 4-Vectors transform correctly under Lorentz transformations. Equations written using 4-vectors are automatically relativistic. This is called the manifestly covariant form of equations. 4-Vectors form a very important part of the formalism of special relativity.

Einstein's Postulate: The Constancy of the Speed of Light

Einstein's postulate that the speed of light is a constant comes out as a natural consequence of the Minkowski formulation.

Proposition 1:

When an object is traveling at c in a certain reference frame, the spacetime interval is zero.

Proof:

The spacetime interval between the origin-event $(0, 0, 0, 0)$ and an event (x, y, z, t) is,

$$s^2 = x^2 + y^2 + z^2 - (ct)^2.$$

The distance travelled by an object moving at velocity v for t seconds is:

$$\sqrt{x^2 + y^2 + z^2} = vt$$

giving,

$$s^2 = (vt)^2 - (ct)^2.$$

Since the velocity v equals c we have,

$$s^2 = (ct)^2 - (ct)^2.$$

Hence the spacetime interval between the events of departure and arrival is given by,

$$s^2 = 0$$

Proposition 2:

An object traveling at c in one reference frame is traveling at c in all reference frames.

Proof:

Let the object move with velocity v when observed from a different reference frame. A change in reference frame corresponds to a rotation in M. Since the spacetime interval must be conserved under rotation, the spacetime interval must be the same in all reference frames. In proposition 1 we showed it to be zero in one reference frame, hence it must be zero in all other reference frames. We get that,

$$(vt)^2 - (ct)^2 = 0$$

which implies,

$$|v| = c.$$

The paths of light rays have a zero spacetime interval, and hence all observers will obtain the same value for the speed of light. Therefore, when assuming that the universe has four dimensions that are related by Minkowski's formula, the speed of light appears as a constant, and does not need to be assumed (postulated) to be constant as in Einstein's original approach to special relativity.

Clock Delays and Rod Contractions

Another consequence of the invariance of the spacetime interval is that clocks will appear to go slower on objects that are moving relative to you. This is very similar to how the 2D projection of a line rotated into the third-dimension appears to get shorter. Length is not conserved simply because we are ignoring one of the dimensions. Let us return to the example of John and Bill.

John observes the length of Bill's spacetime interval as:

$$s^2 = (vt)^2 - (ct)^2$$

whereas Bill doesn't think he has traveled in space, so writes:

$$s^2 = (0)^2 - (cT)^2$$

The spacetime interval, s^2, is invariant. It has the same value for all observers, no matter who measures it or how they are moving in a straight line. This means that Bill's spacetime interval equals John's observation of Bill's spacetime interval so:

$$(0)^2 - (cT)^2 = (vt)^2 - (ct)^2$$

and

$$-(cT)^2 = (vt)^2 - (ct)^2$$

hence

$$t = \frac{T}{\sqrt{1 - \frac{v^2}{c^2}}}.$$

So, if John sees a clock that is at rest in Bill's frame record one second, John will find that his own clock measures between these same ticks an interval *t,* called coordinate time, which is greater than one second. It is said that clocks in motion slow down, relative to those on observers at rest. This is known as "relativistic time dilation of a moving clock." The time that is measured in the rest frame of the clock (in Bill's frame) is called the proper time of the clock.

In special relativity, therefore, changes in reference frame affect time also. Time is no longer absolute. There is no universally correct clock, time runs at different rates for different observers.

Similarly it can be shown that John will also observe measuring rods at rest on Bill's planet to be shorter in the direction of motion than his own measuring rods. This is a prediction known as "relativistic length contraction of a moving rod." If the length of a rod at rest on Bill's planet is X, then we call this quantity the proper length of the rod. The length x of that same rod as measured on John's planet, is called coordinate length, and given by,

$$x = X\sqrt{1 - \frac{v^2}{c^2}}.$$

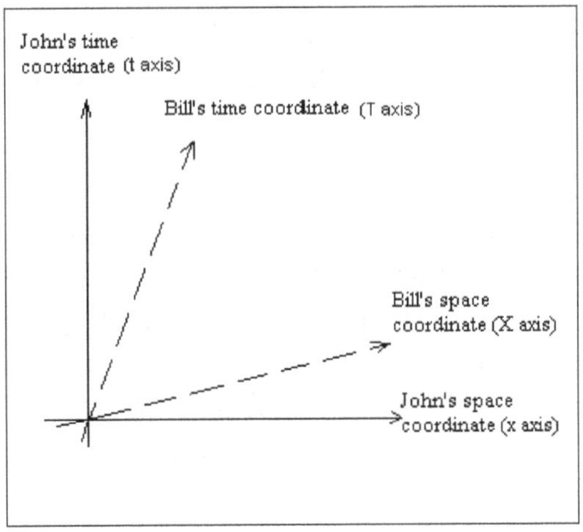

How Bill's coordinates appear to John at the instant they pass each other.

These two equations can be combined to obtain the general form of the Lorentz transformation in one spatial dimension:

$$T = \gamma\left(t - \frac{vx}{c^2}\right)$$
$$X = \gamma(x - vt)$$

or equivalently:

$$t = \gamma\left(T + \frac{vX}{c^2}\right)$$
$$x = \gamma(X + vT)$$

where the Lorentz factor is given by $\gamma = \frac{1}{\sqrt{1 - v^2 / c^2}}$.

The above formulas for clock delays and length contractions are special cases of the general transformation.

Alternatively, these equations for time dilation and length contraction (here obtained from the invariance of the spacetime interval), can be *obtained* directly *from* the Lorentz transformation by setting X = 0 for time dilation, meaning that the clock is at rest in Bill's frame, or by setting t = 0 for length contraction, meaning that John must measure the distances to the end points of the moving rod at the same time.

A consequence of the Lorentz transformations is the modified velocity-addition formula:

$$s = \frac{v + u}{1 + (v/c)(u/c)}.$$

Simultaneity and Clock Desynchronization

Rather counter-intuitively, special relativity suggests that when 'at rest' we are actually moving through time at the speed of light. As we speed up in space we slow down in time. At the speed of light in space, time slows down to zero. This is a rotation of the time axis into the space axis. We observe that the object speeding by relativistically as having its time axis not at a right angle.

The consequence of this in Minkowski's spacetime is that clocks will appear to be out of phase with each other along the length of a moving object. This means that if one observer sets up a line of clocks that are all synchronized so they all read the same time, then another observer who is moving along the line at high speed will see the clocks all reading different times. This means that observers who are moving relative to each other see different events as simultaneous. This effect is known as "Relativistic Phase" or the "Relativity of Simultaneity." Relativistic phase is often overlooked by students of special relativity, but if it is understood, then phenomena such as the twin paradox are easier to understand.

Observers have a set of simultaneous events around them that they regard as composing the present instant. The relativity of simultaneity results in observers who are moving relative to each other having different sets of events in their present instant.

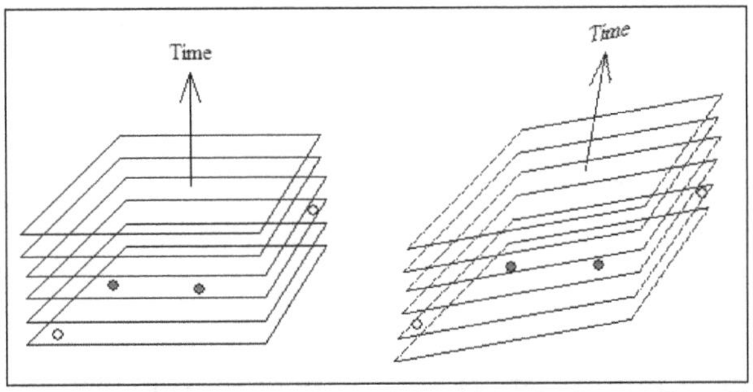

The "plane of simultaneity" or "surface of simultaneity" contains all those events that happen at the same instant for a given observer. Events that are simultaneous for one observer are not simultaneous for another observer in relative motion.

The net effect of the four-dimensional universe is that observers who are in motion relative to you seem to have time coordinates that lean over in the direction of motion, and consider things to be simultaneous that are not simultaneous for you. Spatial lengths in the direction of travel are shortened, because they tip upwards and downwards, relative to the time axis in the direction of travel, akin to a rotation out of three-dimensional space.

Great care is needed when interpreting spacetime diagrams. Diagrams present data in two dimensions, and cannot show faithfully how, for instance, a zero length spacetime interval appears.

Mass Velocity Relationship

$E = mc^{22}$ where m stands for rest mass (invariant mass), applies most simply to single particles with no net momentum. But it also applies to ordinary objects composed of many particles so long as the particles are moving in different directions so the total momentum is zero. The mass of the object includes contributions from heat and sound, chemical binding energies and trapped radiation. Familiar examples are a tank of gas, or a hot bowl of soup. The kinetic energy of their particles, the heat motion and radiation, contribute to their weight on a scale according to $E = mc^{22}$.

The formula is the special case of the relativistic energy-momentum relationship:

$$E^2 - (pc)^2 = (mc^2)^2$$

This equation gives the rest mass of a system which has an arbitrary amount of momentum and energy. The interpretation of this equation is that the rest mass is the relativistic length of the energy-momentum four-vector.

If the equation $E = mc^2$ is used with the rest mass of the object, the E given by the equation will be the rest energy of the object, and will change with according to the object's internal energy, heat and sound and chemical binding energies, but will not change with the object's overall motion.

If the equation $E = mc^2$ is used with the relativistic mass of the object, the energy will be the total energy of the object, which is conserved in collisions with other fast moving objects.

In developing special relativity, Einstein found that the total energy of a moving body is,

$$E = \frac{m_0 c^2}{\sqrt{1 - \frac{v^2}{c^2}}},$$

with v the velocity.

For small velocities, this reduces to,

$$E = m_0 c^2 + \frac{1}{2} m_0 v^2 + ...$$

Which includes the newtonian kinetic energy, as expected, but also an enormous constant term, which is not zero when the object isn't moving.

The total momentum is:

$$P = \frac{m_0 v}{\sqrt{1 - \frac{v^2}{c^2}}}.$$

The ratio of the momentum to the velocity is the relativistic mass, and this ratio is equal to the total energy times c^2. The energy and relativistic mass are always related by the famous formula.

While this is suggestive, it does not immediately imply that the energy and mass are equivalent because the energy can always be redefined by adding or subtracting a constant. So it is possible to subtract the $m_0 c^2$ from the expression for E and this is also a valid conserved quantity, although an ugly one. Einstein needed to know whether the rest-mass of the object is really an energy, or whether the constant term was just a mathematical convenience with no physical meaning.

In order to see if the $m_0 c^2$ is physically significant, Einstein considered processes of emission and absorption. He needed to establish that an object loses mass when it emits energy. He did this by analyzing two photon emission in two different frames.

After Einstein first made his proposal, it became clear that the word mass can have two different meanings. The rest mass is what Einstein called m, but others defined the *relativistic mass* as:

$$m_{rel} = \frac{m_0}{\sqrt{1 - \frac{v^2}{c^2}}}.$$

This mass is the ratio of momentum to velocity, and it is also the relativistic energy divided by c^2. So the equation $E = m_{rel} c^2$ holds for moving objects. When the velocity is small, the relativistic mass and the rest mass are almost exactly the same.

$E = mc^2$ either means $E = m_0 c^2$ for an object at rest, or $E = m_{rel} c^2$ when the object is moving.

Einstein's original papers treated *m* as what would now be called the *rest mass* and some claim that he did not like the idea of "relativistic mass." When modern physicists say "mass, " they are usually talking about rest mass, since if they meant "relativistic mass, " they would just say "energy."

We can rewrite the expression for the energy as a Taylor series:

$$E = m_0 c^2 \left[1 + \frac{1}{2}\left(\frac{v}{c}\right)^2 + \frac{3}{8}\left(\frac{v}{c}\right)^4 + \frac{5}{16}\left(\frac{v}{c}\right)^6 + \ldots \right].$$

For speeds much smaller than the speed of light, higher-order terms in this expression get smaller and smaller because v / c is small. For low speeds we can ignore all but the first two terms:

$$E \approx m_0 c^2 + \frac{1}{2} m_0 v^2.$$

The total energy is a sum of the rest energy and the Newtonian kinetic energy.

The classical energy equation ignores both the $m_0 c^2$ part, and the high-speed corrections. This is appropriate, because all the high order corrections are small. Since only *changes* in energy affect the behavior of objects, whether we include the $m_0 c^2$ part makes no difference, since it is constant. For the same reason, it is possible to subtract the rest energy from the total energy in relativity. In order to see if the rest energy has any physical meaning, it is essential to consider emission and absorption of energy in different frames.

The higher-order terms are extra correction to Newtonian mechanics which become important at higher speeds. The Newtonian equation is only a low speed approximation, but an extraordinarily good one. All of the calculations used in putting astronauts on the moon, for example, could have been done using Newton's equations without any of the higher order corrections.

Mass-energy Equivalence: Sunlight and Atom Bombs

Einstein showed that mass is simply another form of energy. The energy equivalent of rest mass m is $E = mc^2$. This equivalence implies that mass should be interconvertible with other forms of energy. This is the basic principle behind atom bombs and production of energy in nuclear reactors and stars (like Sun).

The standard model of the structure of matter has it that most of the 'mass' of the atom is in the atomic nucleus, and that most of this nuclear mass is in the intense field of light-like gluons swathing the quarks. Most of what is called the mass of an object is thus already in the form of energy, the energy of the quantum color field that confines the quarks.

The sun, for instance, fuels its prodigious output of energy by converting each second 600 billion kilograms of hydrogen-1 (single protons) into 595.2 billion kilograms of helium-4 (2 protons combined with 2 neutrons)—the 4.2 billion kilogram difference is the energy which the sun radiates into space each second. The sun, it is estimated, will continue to turn 4.2 billion kilos of mass into energy for the next 5 billion years or so before leaving the main sequence.

The atomic bombs that ended the Second World War, in comparison, converted about a thirtieth of an ounce of mass into energy.

The energy involved in chemical reactions is so small, however, that the conservation of mass is an excellent approximation.

General Relativity: A Peek Forward

Unlike Newton's laws of motion, relativity is not based upon dynamical postulates. It does not assume anything about motion or forces. Rather, it deals with the fundamental nature of spacetime. It is concerned with describing the geometry of the backdrop on which all dynamical phenomena take place. In a sense therefore, it is a meta-theory, a theory that lays out a structure that all other theories must follow. In truth, Special relativity is only a special case. It assumes that spacetime is flat. That is, it assumes that the structure of Minkowski space and the Minkowski metric tensor is constant throughout. In General relativity, Einstein showed that this is not true. The structure of spacetime is modified by the presence of matter. Specifically, the distance formula given above is no longer generally valid except in space free from mass. However, just like a curved surface can be considered flat in the infinitesimal limit of calculus, a curved spacetime can be considered flat at a small scale. This means that the Minkowski metric written in the differential form is generally valid.

$$ds^2 = dx^2 + dy^2 + dz^2 - c^2dt^2$$

One says that the Minkowski metric is valid locally, but it fails to give a measure of distance over extended distances. It is not valid globally. In fact, in general relativity the global metric itself becomes dependent on the mass distribution and varies through space. The central problem of general relativity is to solve the famous Einstein field equations for a given mass distribution and find the distance formula that applies in that particular case. Minkowski's spacetime formulation was the conceptual stepping stone to general relativity. His fundamentally new outlook allowed not only the development of general relativity, but also to some extent quantum field theories.

General Relativity

General relativity, part of the wide-ranging physical theory of relativity formed by the German-born physicist Albert Einstein. It was conceived by Einstein in 1916. General relativity is concerned with gravity, one of the fundamental forces in the universe. Gravity defines macroscopic behaviour, and so general relativity describes large-scale physical phenomena.

General relativity follows from Einstein's principle of equivalence: on a local scale it is impossible to distinguish between physical effects due to gravity and those due to acceleration. Gravity is treated as a geometric phenomenon that arises from the curvature of space-time. The solution of the fieldequations that describe general relativity can yield answers to different physical situations, such as planetary dynamics, the birth and death of stars, black holes, and the evolution of the universe. General relativity has been experimentally verified by observations of gravitational lenses, the orbit of the planet Mercury, the dilation of time in Earth's gravitational field, and gravitational waves from merging black holes.

German-Swiss-American physicist Albert Einstein showed that time and distance can be measured to have different values depending on their relative speed and direction. However, special relativity only applies to objects moving at a constant velocity. If an object accelerates, then the theory breaks down.

Einstein's theory of general relativity applies special relativity to the concept of acceleration.

English natural philosopher Isaac Newton had already shown the equivalence between gravitational acceleration and ordinary acceleration, and so general relativity unites theories of space and time with theories of light and matter.

The effects caused by acceleration and the effects caused by gravity are the same.

Gravity can be equated with acceleration because the effects of a gravitational field can be produced by accelerating. In a closed room, there's no experiment a person can conduct that will tell them if objects are falling because they are experiencing a gravitational force, or because they are in deep space, accelerating at the same rate.

Einstein realised that a gravitational field would cause objects to accelerate if Minkowski's spacetime were curved. This acceleration is similar to when an object rolls down a curved surface, like a hill. In this case, however, the amount of acceleration is higher for objects with a higher mass because a higher mass causes a deeper, longer curve.

This means that objects accelerate because of the curvature of spacetime, and we call this effect gravity.

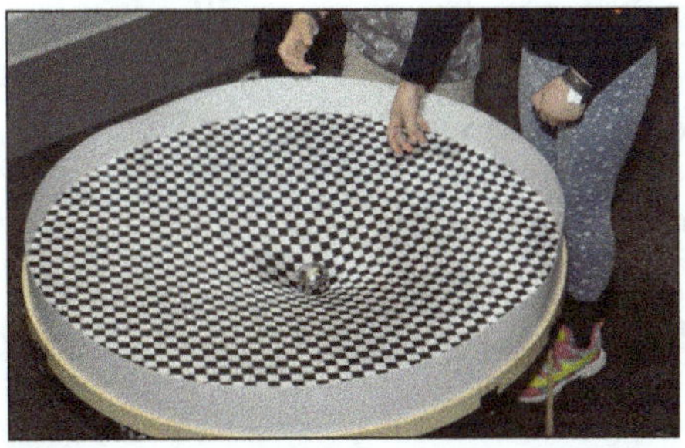

Massive objects, like stars, curve spacetime in a similar way to how massive objects, like a heavy ball, will curve a stretched-out piece of fabric. A lighter object, like a coin, may spin around the curve created by the ball, just as a lighter object like a planet may spin around the curve created by the Sun.

The moon spins around the curve - or gravitational well - created by the Earth.

The relationship between spacetime and mass was summarised by American physicist John Wheeler as follows:

"Spacetime tells matter how to move; matter tells spacetime how to curve".

Einstein showed how the curvature of spacetime is related to the distribution of matter using equations developed by German mathematician Bernhard Riemann in the 1850s. Einstein showed that:

$$G_{\mu\nu} = R_{\mu\nu} - \frac{1}{2} R g_{\mu\nu}$$

$$G_{\mu\nu} = \frac{8\pi G}{c^4} T_{\mu\nu}$$

Here, $G_{\mu\nu}$ is the Einstein tensor, which describes the curvature of spacetime. $R_{\mu\nu}$ is the Ricci tensor, which describes the relationship between Euclidean and non-Euclidean geometry. $g_{\mu\nu}$ is the metric tensor for Minkowski space, and describes all the intrinsic properties of the spacetime manifold, including time periods, distances, and volumes, as well as the curvature, and finally, $T_{\mu\nu}$ is the energy-momentum tensor, which describes the distribution of matter. The energy-momentum tensor is the source of the gravitational field, just as mass is the source of the gravitational field in Newton's equations.

A tensor is like a vector, which contains two properties. Velocity, for example, is a vector as it represents speed in a given direction. A tensor contains more than two properties, which may be written in a matrix - numbers or symbols that are arranged in rows and columns.

General relativity shows that observers in any frame will agree on how spacetime is curved by objects, and hence their gravitational field, whether they are moving relative the object or not. This means that the curvature of space, and hence the force of gravity is invariant.

Light moves through curved spacetime, taking the shortest possible path, however the shortest path across a curved surface is not necessarily a straight line. The shortest path, which may be curved, is known as a geodesic. This meant general relativity predicted that the path of light is bent by heavy objects, like the Sun.

The Cosmological Principle

In order to apply his theories to the universe as a whole, Einstein applied the cosmological principle. This assumes that the universe is homogeneous and isotropic when averaged over very large scales.

Homogeneity assumes that our observations are representational of the whole universe, and isotropy means that the universe is the same in whichever direction we look. These ideas were formulised by British astronomer Edward Milne in the 1930s, and verified by NASA's WMAP (Wilkinson Microwave Anisotropy Probe), which launched in 2001.

Confirmation of General Relativity

British physicist Arthur Eddington confirmed general relativity after the 1919 solar eclipse. Eddington knew that if mass curves spacetime, then light would travel in a curved path as it approaches a massive object like the Sun. This means that if the glare of the Sun were blocked out, like it is during an eclipse, he would be able to see stars that should be behind it. Einstein's theory predicted that the gravitational force of the Sun can cause starlight to deflect by up to 1.75 arc seconds (0.0005°), and Eddington confirmed this.

Eddington's photograph of the 1919 eclipse with dashed lines showing the positions of stars. The stars appeared to have moved from their usual relative positions because their light is bent around the gravitational well of the Sun.

The light from distant galaxies is affected by the curvature of spacetime caused by a massive cluster of galaxies - Abell 2218. This has caused images of the galaxies to be enlarged and duplicated, and is known as gravitational lensing.

Consequences of General Relativity

Precession of Mercury

The curvature of spacetime forces planets to orbit in open ellipses that are rotating. This effect is more noticeable the closer a planet is to the Sun, and so this explained why these effects were first observed in Mercury's orbit.

Gravitational Lensing

The curvature of spacetime means that the path of light is deflected around massive objects. This effect is known as gravitational lensing, and it can affect the shape of an event's light cone, allowing light to travel into previously forbidden regions. Gravitational lensing magnifies and brightens an image, and it's also possible for the same image to be projected more than once, as light is deflected in different directions. Eddington showed that light is deflected around the Sun in 1919.

Swiss astrophysicist Fritz Zwicky first considered using galaxy clusters as gravitational lenses in 1937.These effects were first observed by astronomers Dennis Walsh, Robert Carswell, and Ray Weymann, at the Kitt Peak National Observatory in the United States, in 1979.

Gravitational Waves

As massive objects move through spacetime, they will change its curvature, and this change will dissipate from the objects at the speed of light, creating 'ripples' in spacetime, like ripples on a pond. These ripples are known as gravitational waves. Scientists in the LIGO (Laser Interferometer Gravitational-Wave Observatory) Scientific Collaboration and the Virgo Collaboration discovered evidence for gravitational waves from a pair of merging black holes in 2016.

An artist's impression of a binary star system containing two neutron stars. The neutron stars create curves in spacetime as they orbit each other, and these dissipate at the speed of light, creating waves.

Black Holes

Black holes occur when an object becomes so massive that even light cannot escape its gravitational well. We now know that supermassive black holes reside at the centre of most galaxies, including our own. These are millions of times the mass of the Sun.

The more massive the object, the more it curves spacetime, and the steeper the gravitational well. When an object is so massive that even light, which has no mass, cannot escape the well, the object is known as a 'black hole'.

Time Travel

If mass curves space, then it will also curve time, and in 1949, Austrian mathematician Kurt Gödel suggested that time travel is possible because closed timelike curves can occur. These are regions where time is curved so that an observer would find themselves in an earlier time. Time travel would violate our notion of causality, and so there's still much debate over whether it's really possible.

Gravitational Time Dilation

By 1907, Einstein had already predicted that gravity would create an effect similar to the Doppler effect. Einstein predicted that from the point of view of a stationary observer, light will appear to be red shifted as it moves towards a strong gravitational force. This change corresponds to the fact that time appears to run slower for observers in a stronger gravitational field.

'Proper' time, or invariant time (t), is related to the relative time measured by something moving towards or away from a gravitational field (t_{rel}) via:

$$\Delta t_{rel} = \frac{\Delta t}{\sqrt{1 - \frac{2Gm}{rc^2}}}$$

Gravitational time dilation was first observed in 1971, when American physicists Joseph Hafele and Richard Keating flew four atomic clocks twice around the world on a commercial plane, and then compared their time to clocks left on Earth.

All satellite navigation systems, including Global Positioning Systems (GPS), have to take the effects of gravitational time dilation into account in order to be accurate.

Gravitational time dilation means that light appears redder as it moves towards a gravitational field.

Absolute Space

Einstein suggested that general relativity disproves Newton's idea that space and time exist independently of matter and energy, as a type of substance, stating that general relativity "takes away the last remnants of physical objectivity from space and time".

This is because, in general relativity, spacetime is no longer considered a background in which events take place. Instead, spacetime is affected by what it contains. All observable events are described by the metric tensor, but there's no unique coordinate system. The choice is arbitrary. This means that the actual shape of spacetime can be described in a number of different ways that all give the same observable results. There's no experiment that can distinguish between them. If spacetime is a real substance, then it would presumably have to pick one of these shapes, and this leads to the 'hole' argument.

The hole argument shows that if you accept that coordinate systems do represent the true nature of spacetime, then there's no reason why there couldn't be a 'hole', where spacetime suddenly follows another coordinate system. There's no way to determine when a hole will appear, since both systems are observationally identical, and so proponents of absolute spacetime must accept that general relativity possesses the same objective uncertainty as quantum mechanics. If we remove the idea that spacetime is absolute, then we do not face this problem.

This does not necessarily address Newton's argument that absolute space can be derived from the fact that acceleration is absolute; we can always determine if we are accelerating. We cannot feel the effects of the movement of the Earth, for example, but we can prove that we are moving, and so some continue to argue that spacetime must be absolute.

A Non-static Universe

For Einstein, the most startling consequence of general relativity was that it shows that the shape of the universe depends on its energy density. This means that if all the mass in the universe is not sufficiently distributed, then everything will eventually fall in on itself, and the universe will end in a 'big crunch'. Einstein devised the cosmological constant in order to counterbalance this force and create a static universe.

In 1922, Russian mathematician Alexander Friedmann showed that it's also possible that the universe is expanding. This could occur if it begun at an extremely high density and temperature. If this were the case, and if mass is distributed too sparsely, then matter will continue to move apart forever.

American astronomer Edwin Hubble proved that the universe is expanding in 1929 . It's now thought that the universe will expand forever due to dark energy, which was discovered in 1998.

Advanced Concepts

Causal Structure and Global Geometry

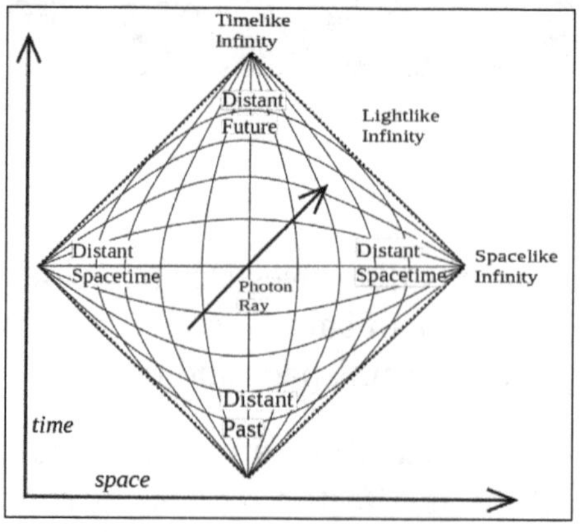

Penrose–Carter diagram of an infinite Minkowski universe.

In general relativity, no material body can catch up with or overtake a light pulse. No influence from an event *A* can reach any other location *X*before light sent out at *A* to *X*. In consequence, an exploration of all light worldlines (null geodesics) yields key information about the spacetime's causal structure. This structure can be displayed using Penrose–Carter diagrams in which infinitely large regions of space and infinite time intervals are shrunk ("compactified") so as to fit onto a finite map, while light still travels along diagonals as in standard spacetime diagrams.

Aware of the importance of causal structure, Roger Penrose and others developed what is known as global geometry. In global geometry, the object of study is not one particular solution (or family of solutions) to Einstein's equations. Rather, relations that hold true for all geodesics, such as the Raychaudhuri equation, and additional non-specific assumptions about the nature of matter (usually in the form of energy conditions) are used to derive general results.

Horizons

Using global geometry, some spacetimes can be shown to contain boundaries called horizons, which demarcate one region from the rest of spacetime. The best-known examples are black holes: if mass is compressed into a sufficiently compact region of space (as specified in the hoop conjecture, the relevant length scale is the Schwarzschild radius), no light from inside can escape to the outside. Since no object can overtake a light pulse, all interior matter is imprisoned as well. Passage from the exterior to the interior is still possible, showing that the boundary, the black hole's *horizon*, is not a physical barrier.

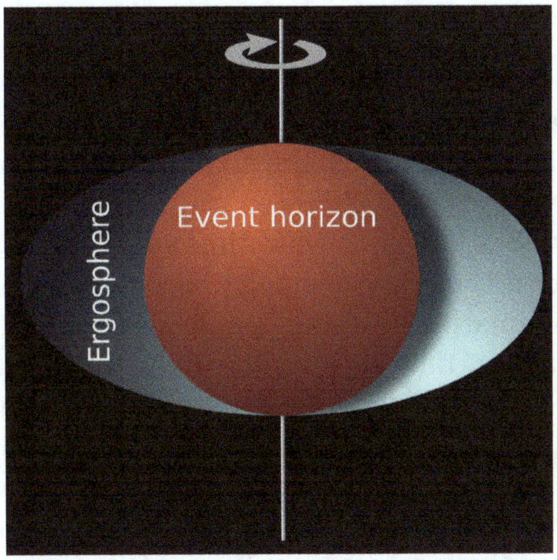

The ergosphere of a rotating black hole, which plays a key role when it comes to extracting energy from such a black hole.

Early studies of black holes relied on explicit solutions of Einstein's equations, notably the spherically symmetric Schwarzschild solution (used to describe a static black hole) and the axisymmetric Kerr solution (used to describe a rotating, stationary black hole, and introducing interesting features such as the ergosphere). Using global geometry, later studies have revealed more general properties of black holes. In the long run, they are rather simple objects characterized by eleven parameters specifying energy, linear momentum, angular momentum, location at a specified time and electric charge. This is stated by the black hole uniqueness theorems: "black holes have no hair", that is, no distinguishing marks like the hairstyles of humans. Irrespective of the complexity of a gravitating object collapsing to form a black hole, the object that results (having emitted gravitational waves) is very simple.

Even more remarkably, there is a general set of laws known as black hole mechanics, which is analogous to the laws of thermodynamics. For instance, by the second law of black hole mechanics, the area of the event horizon of a general black hole will never decrease with time, analogous to the entropy of a thermodynamic system. This limits the energy that can be extracted by classical means from a rotating black hole (e.g. by the Penrose process). There is strong evidence that the laws of black hole mechanics are, in fact, a subset of the laws of thermodynamics, and that the black hole area is proportional to its entropy. This leads to a modification of the original laws of black hole mechanics: for instance, as the second law of black hole mechanics becomes part of the second law of thermodynamics, it is possible for black hole area to decrease—as long as other processes ensure that, overall, entropy increases. As thermodynamical objects with non-zero temperature, black holes should emit thermal radiation. Semi-classical calculations indicate that indeed they do, with the surface gravity playing the role of temperature in Planck's law. This radiation is known as Hawking radiation.

There are other types of horizons. In an expanding universe, an observer may find that some regions of the past cannot be observed ("particle horizon"), and some regions of the future cannot be influenced (event horizon). Even in flat Minkowski space, when described by an accelerated observer (Rindler space), there will be horizons associated with a semi-classical radiation known as Unruh radiation.

Singularities

Another general feature of general relativity is the appearance of spacetime boundaries known as singularities. Spacetime can be explored by following up on timelike and lightlike geodesics—all possible ways that light and particles in free fall can travel. But some solutions of Einstein's equations have "ragged edges"—regions known as spacetime singularities, where the paths of light and falling particles come to an abrupt end, and geometry becomes ill-defined. In the more interesting cases, these are "curvature singularities", where geometrical quantities characterizing spacetime curvature, such as the Ricci scalar, take on infinite values. Well-known examples of spacetimes with future singularities—where worldlines end—are the Schwarzschild solution, which describes a singularity inside an eternal static black hole, or the Kerr solution with its ring-shaped singularity inside an eternal rotating black hole. The Friedmann–Lemaître–Robertson–Walker solutions and other spacetimes describing universes have past singularities on which worldlines begin, namely Big Bang singularities, and some have future singularities (Big Crunch) as well.

Given that these examples are all highly symmetric—and thus simplified—it is tempting to conclude that the occurrence of singularities is an artifact of idealization. The famous singularity theorems, proved using the methods of global geometry, say otherwise: singularities are a generic feature of general relativity, and unavoidable once the collapse of an object with realistic matter properties has proceeded beyond a certain stage and also at the beginning of a wide class of expanding universes. However, the theorems say little about the properties of singularities, and much of current research is devoted to characterizing these entities' generic structure (hypothesized e.g. by the BKL conjecture).The cosmic censorship hypothesis states that all realistic future singularities (no perfect symmetries, matter with realistic properties) are safely hidden away behind a horizon, and thus invisible to all distant observers. While no formal proof yet exists, numerical simulations offer supporting evidence of its validity.

Evolution Equations

Each solution of Einstein's equation encompasses the whole history of a universe — it is not just some snapshot of how things are, but a whole, possibly matter-filled, spacetime. It describes the state of matter and geometry everywhere and at every moment in that particular universe. Due to its general covariance, Einstein's theory is not sufficient by itself to determine the time evolution of the metric tensor. It must be combined with a coordinate condition, which is analogous to gauge fixing in other field theories.

To understand Einstein's equations as partial differential equations, it is helpful to formulate them in a way that describes the evolution of the universe over time. This is done in "3+1" formulations, where spacetime is split into three space dimensions and one time dimension. The best-known example is the ADM formalism. These decompositions show that the spacetime evolution equations of general relativity are well-behaved: solutions always exist, and are uniquely defined, once suitable initial conditions have been specified.Such formulations of Einstein's field equations are the basis of numerical relativity.

Global and Quasi-local Quantities

The notion of evolution equations is intimately tied in with another aspect of general relativistic physics. In Einstein's theory, it turns out to be impossible to find a general definition for a seemingly simple property such as a system's total mass (or energy). The main reason is that the gravitational field—like any physical field—must be ascribed a certain energy, but that it proves to be fundamentally impossible to localize that energy.

Nevertheless, there are possibilities to define a system's total mass, either using a hypothetical "infinitely distant observer" (ADM mass) or suitable symmetries (Komar mass). If one excludes from the system's total mass the energy being carried away to infinity by gravitational waves, the result is the Bondi mass at null infinity. Just as in classical physics, it can be shown that these masses are positive. Corresponding global definitions exist for momentum and angular momentum. There have also been a number of attempts to define *quasi-local* quantities, such as the mass of an isolated system formulated using only quantities defined within a finite region of space containing that system. The hope is to obtain a quantity useful for general statements about isolated systems, such as a more precise formulation of the hoop conjecture.

Relationship with Quantum Theory

If general relativity were considered to be one of the two pillars of modern physics, then quantum theory, the basis of understanding matter from elementary particles to solid state physics, would be the other. However, how to reconcile quantum theory with general relativity is still an open question.

Quantum Field Theory in Curved Spacetime

Ordinary quantum field theories, which form the basis of modern elementary particle physics, are defined in flat Minkowski space, which is an excellent approximation when it comes to describing the behavior of microscopic particles in weak gravitational fields like those found on Earth. In order to describe situations in which gravity is strong enough to influence (quantum) matter, yet not strong enough to require quantization itself, physicists have formulated quantum field theories in curved spacetime. These theories rely on general relativity to describe a curved background spacetime, and define a generalized quantum field theory to describe the behavior of quantum matter within that spacetime.Using this formalism, it can be shown that black holes emit a blackbody spectrum of particles known as Hawking radiation leading to the possibility that they evaporate over time.As briefly mentioned above, this radiation plays an important role for the thermodynamics of black holes.

Quantum Gravity

The demand for consistency between a quantum description of matter and a geometric description of spacetime, as well as the appearance of singularities (where curvature length scales become microscopic), indicate the need for a full theory of quantum gravity: for an adequate description of the interior of black holes, and of the very early universe, a theory is required in which gravity and the associated geometry of spacetime are described in the language of quantum physics. Despite major efforts, no complete and consistent theory of quantum gravity is currently known, even though a number of promising candidates exist.

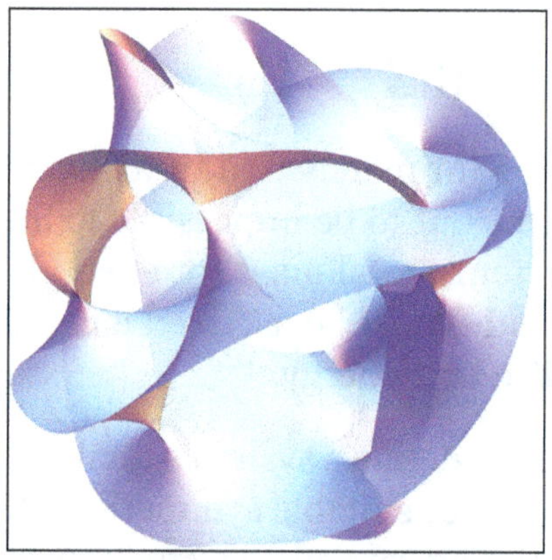

Projection of a Calabi–Yau manifold, one of the ways of compactifying the extra dimensions posited by string theory.

Attempts to generalize ordinary quantum field theories, used in elementary particle physics to describe fundamental interactions, so as to include gravity have led to serious problems. Some have argued that at low energies, this approach proves successful, in that it results in an acceptable effective (quantum) field theory of gravity. At very high energies, however, the perturbative results are badly divergent and lead to models devoid of predictive power ("perturbative non-renormalizability").

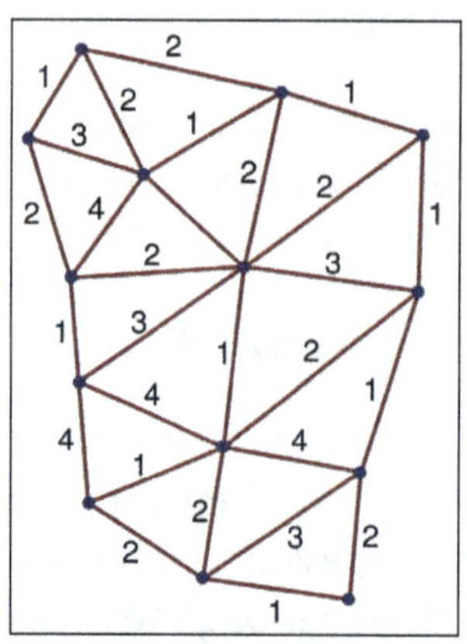

Simple spin network of the type used in loop quantum gravity.

One attempt to overcome these limitations is string theory, a quantum theory not of point particles, but of minute one-dimensional extended objects. The theory promises to be a unified descriptionof all particles and interactions, including gravity; the price to pay is unusual features such as six extra dimensions of space in addition to the usual three. In what is called the second superstring revolution, it was conjectured that both string theory and a unification of general relativity and supersymmetry known as supergravity form part of a hypothesized eleven-dimensional model known as M-theory, which would constitute a uniquely defined and consistent theory of quantum gravity.

Another approach starts with the canonical quantization procedures of quantum theory. Using the initial-value-formulation of general relativity (cf. evolution equations above), the result is the Wheeler–deWitt equation (an analogue of the Schrödinger equation) which, regrettably, turns out to be ill-defined without a proper ultraviolet (lattice) cutoff. However, with the introduction of what are now known as Ashtekar variables, this leads to a promising model known as loop quantum gravity. Space is represented by a web-like structure called a spin network, evolving over time in discrete steps.

Depending on which features of general relativity and quantum theory are accepted unchanged, and on what level changes are introduced, there are numerous other attempts to arrive at a viable theory of quantum gravity, some examples being the lattice theory of gravity based on the Feynman Path Integral approach and Regge Calculus, dynamical triangulations, causal sets, twistor models or the path integral based models of quantum cosmology.

All candidate theories still have major formal and conceptual problems to overcome. They also face the common problem that, as yet, there is no way to put quantum gravity predictions to experimental tests (and thus to decide between the candidates where their predictions vary), although there is hope for this to change as future data from cosmological observations and particle physics experiments becomes available.

Current Status

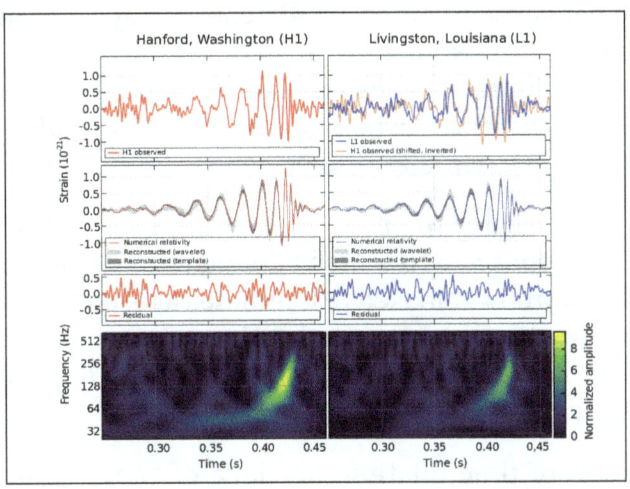

Observation of gravitational waves from binary black hole merger GW150914.

General relativity has emerged as a highly successful model of gravitation and cosmology, which has so far passed many unambiguous observational and experimental tests. However, there are strong indications the theory is incomplete. The problem of quantum gravity and the question of the reality of spacetime singularities remain open. Observational data that is taken as evidence for dark energy and dark matter could indicate the need for new physics. Even taken as is, general relativity is rich with possibilities for further exploration. Mathematical relativists seek to understand the nature of singularities and the fundamental properties of Einstein's equations, while numerical relativists run increasingly powerful computer simulations (such as those describing merging black holes). In February 2016, it was announced that the existence of gravitational waves was directly detected by the Advanced LIGO team on September 14, 2015. A century after its introduction, general relativity remains a highly active area of research.

Relativistic Space-time

The concept of relativistic spacetime is fundamental to the structure of modern physics. The measurement of time is no longer absolute, but is, according to special relativity theory, path-dependent. In Einstein's terms, an observer at rest in one reference frame measures his own *proper time ("eigen-zeit")* on a clock, while observers moving relative to him can calculate his proper time using their measurements of both distance and time. From Albert Einstein's general relativity theory—the relativistic theory of gravitation—it is possible to derive three standard models of the Universe that allow for both finite and infinite temporal developments, all starting from an initial singularity. In deciding between these models, quantum mechanics—the modern (nonclassical) theory of matter—must be taken into account. Newton's conception of time now remains valid only locally and approximately for velocities considerably lower than the velocity of light.

Einstein's former mathematics professor, Hermann Minkowski, was perhaps the first to note this effect (and perhaps understood it even better than Einstein himself), and it was he who coined the phrase "space-time" to describe the interchangeability of the four dimensions. In 1908, Minkowski offered a useful analogy to help explain how four-dimensional space-time can appear differently to two observers in our normal three-dimensional space. He described two observers viewing a three-dimensional object from different angles, and noting that, for example, the length and width can appear different from the different viewpoints, due to what we call perspective, even though the object is clearly one and the same in three dimensions.

The idea perhaps becomes even clearer when we consider that our picture of the Moon is actually what the Moon was like 1¼ seconds ago (the time light takes to reach the Earth from the Moon), our picture of the Sun is actually how it looked 8½ minutes ago, and by the time we see an image of Alpha Centauri, our nearest star system, it is already 4.3 years out of date. We can therefore never know what the universe is like at this very instant, and the universe is clearly not a thing that extends just in space, but in space-time.

Due to the relativistic effects of time dilation, our idea of "now" is therefore something of a fictitious concept, one which we as humans have invented for ourselves, but for which nature itself has no real use. Physicists do not regard time as "passing" or "flowing" and time is not a sequence of events which happen: the past and the future are simply there, laid out as part of space-time.

But the malleability and blurring of space and time also has implications for other aspects of physics. Just as Maxwell had shown that the electric and magnetic fields, once considered completely separate entities, were both just part of a single seamless entity known as the electromagnetic field, likewise (although perhaps more difficult to grasp and perhaps more unexpected) energy and mass turn out to be just different faces of the same coin, a connection encapsulated in Einstein's justifiably famous formula, $E = mc^2$.

Relativistic Momentum

Momentum is one of the most important concepts in physics. The broadest form of Newton's second law is stated in terms of momentum. Momentum is conserved whenever the net external force on a system is zero. This makes momentum conservation a fundamental tool for analyzing collisions. All of Work, Energy, and Energy Resources is devoted to momentum, and momentum has been important for many other topics as well, particularly where collisions were involved. We will see that momentum has the same importance in modern physics. Relativistic momentum is conserved, and much of what we know about subatomic structure comes from the analysis of collisions of accelerator-produced relativistic particles.

The first postulate of relativity states that the laws of physics are the same in all inertial frames. Does the law of conservation of momentum survive this requirement at high velocities? The answer is yes, provided that the momentum is defined as follows.

Relativistic momentum is defined in such a way that the conservation of momentum will hold in all inertial frames. Whenever the net external force on a system is zero, relativistic momentum is conserved, just as is the case for classical momentum. This has been verified in numerous experiments.

In Relativistic Energy, the relationship of relativistic momentum to energy is explored. That subject will produce our first inkling that objects without mass may also have momentum.

Classically, momentum is defined by,

$$\vec{p} = m\vec{u}$$

where m and $\vec{u}$ the mass and velocity of an object. What is the correct relativistic definition of momentum? We have the freedom to define momentum any way we like, but to be useful, we would like to define it such that total momentum is conserved in all inertial frames.

Recall the law of momentum conservation. If there are n bodies with momenta $\vec{p}_1, \ldots, \vec{p}_n$, then, in the absence of external forces, the total momentum is given by,

$$\sum \vec{p} = \vec{p}_1 + \ldots + \vec{p}_n$$

cannot change. We would like to define momentum relativistically to preserve conservation of momentum.

We can't use the classical definition Eq. $\vec{p} = m\vec{u}$ because of the way velocity transforms between reference frames. For example, the way u_y transforms in going from S to S ′ depends on u_x. The classical definition of momentum is,

$$\vec{p} = m\vec{u} = m\frac{d\vec{r}}{dt}$$

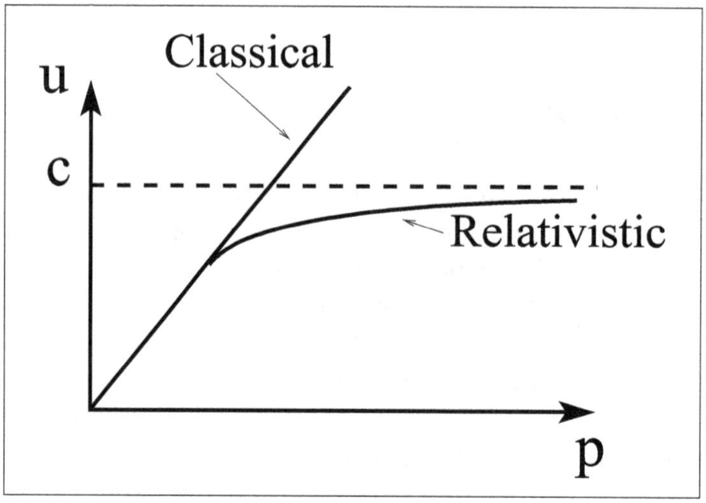

Velocity versus momentum.

The problem is that both $\vec{r}$ and t are subject to the Lorentz transformation and that makes things messy.

The relativistic way to define momentum is,

$$\vec{p} = m\frac{d\vec{r}}{dt_o}$$

where t_o is the proper time, i.e., the time measured in the rest frame of the particle or object. Notice that with this definition,

$$p_y = m\frac{dy}{dt_o}$$

which is invariant when one goes between reference frames that are moving relative to one another along the x–axis. That is because t_o is invariant and y = y ′. Similarly for p_z.

Since dt = γdt_o where $\gamma = 1\sqrt{1-u^2/c^2}$, we can write,

$$\vec{p} = \frac{m\vec{u}}{\sqrt{1-u^2/c^2}} = \gamma m\vec{u}$$

A constant force changes the momentum and hence the velocity. Notice that as u → c, γ and hence p increase without limit, but the speed u never reaches the speed of light. Thus no object can go faster than the speed of light.

We can define a relativistic mass M(u) by,

$$M(u)=\frac{M}{\sqrt{1-u^2/c^2}}=M\gamma$$

calls this a variable mass m_{var}, but it doesn't want to use this. However, it does make it easy to write the momentum in the traditional form,

$$\vec{p}=M(u)\vec{u}$$

Notice from Eq.mentioned before the above one, that if the speed u of an object with nonzero mass is equal to the speed of light, the relativistic mass is infinity. If the object's speed exceeds the speed of light, then the relativistic mass M(u) is imaginary. Clearly, this is absurd and so, once again, the velocity of an object cannot exceed the speed of light. If the object has mass, it cannot travel at the speed of light; its speed must be less than the speed of light.

Relativistic Mass

Relativistic mass refers to mass of a body which change with the speed of the body as this speeds approaches close to speed of light, it increases with velocity and tends to infinity when the velocity approaches the speed of light.

Concept of Relativistic Mass

In special relativity, the expressions for the total energy and momentum of an object can be written as,

$$E=m_0\gamma c\ ,\ \vec{p}=m_0\gamma\vec{v},$$

where $\gamma=\frac{1}{\sqrt{1-v^2/c^2}}$ and m_0 is the invariant mass (or rest mass) of the object. If we introduce the relativistic mass,

$$m=m_0\gamma$$

these expressions turn into,

which many find more familiar since the first is an iconic relation in special relativity and the latter coincides with the classical expression for momentum.

As almost everyone knows, Newton's 2nd law says that,

$$F=ma$$

where F is the force applied, m is the mass, and a is the acceleration felt by the body. In Newtonian mechanics, mass is invariant, but a consequence of special relativity is that nothing can travel faster than the speed of light c. This raises the conundrum of why can't we keep applying a force to a body of mass m, causing it to continue accelerating and to ultimately increase its velocity to one greater than the speed of light?

The answer is that Newton's 2nd law is incomplete. Einstein showed that mass is also a function of velocity, and so we should write,

$$m = \gamma m_o$$

Where $\gamma = \frac{1}{\sqrt{(1 - V^2 / c^2)}}$ is the so-called *Lorentz factor* and m_o is the rest mass (also known as the invariant mass or gravitational mass), the mass an object has when it is at rest relative to the observer. Hence we can argue that, as we approach the speed of light, the applied force goes into changing the mass of the body, rather than accelerating it, leading to a modified version of Newton's 2nd law,

$$F \quad m \; a$$

where both velocity and/or mass change as a force is applied. But, because of the fact that $\gamma \approx 1$ until $V \approx c/2$, very little increase in mass occurs until V has reached appreciable values.

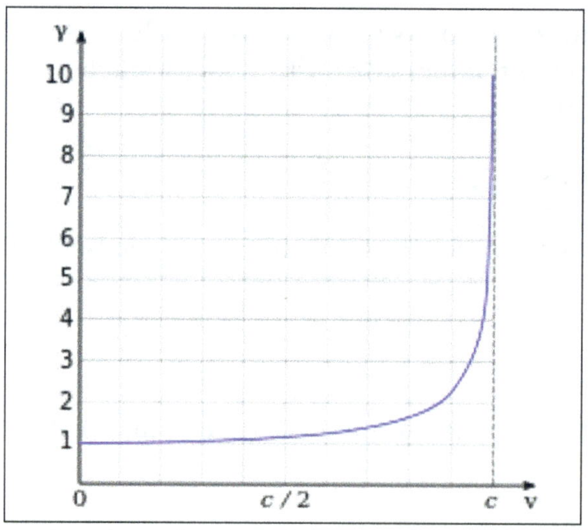

The variation of γ (the Lorentz factor) as a function of the speed V. Until $V \approx c/2$, γ is very close to unity.

Relativistic Energy

The relativistic energy is the way that Einstein showed that the law of conservation of energy is valid relativistically, it means, the law of conservation of energy is valid in all inertial frames in high velocities approaching to the speed of light.

Relativistic energy = rest mass * speed of light squared / squared root [one minus (velocity / speed of light) squared].

The equation is:

$$E = mc^2 / \text{sqrt}\,(1 - v^2 / c^2)$$

Where:

E: relativistic energy

m: rest mass (invariant mass)

v: velocity of the body

c: speed of light

Conservation of energy is one of the most important laws in physics. Not only does energy have many important forms, but each form can be converted to any other. We know that classically the total amount of energy in a system remains constant. Relativistically, energy is still conserved, provided its definition is altered to include the possibility of mass changing to energy, as in the reactions that occur within a nuclear reactor. Relativistic energy is intentionally defined so that it will be conserved in all inertial frames, just as is the case for relativistic momentum. As a consequence, we learn that several fundamental quantities are related in ways not known in classical physics. All of these relationships are verified by experiment and have fundamental consequences. The altered definition of energy contains some of the most fundamental and spectacular new insights into nature found in recent history.

Total Energy and Rest Energy

The first postulate of relativity states that the laws of physics are the same in all inertial frames. Einstein showed that the law of conservation of energy is valid relativistically, if we define energy to include a relativistic factor.

Total Energy

Total energy E is defined to be $E = \gamma mc^2$, where m is mass, c is the speed of light,

$$\gamma = \frac{1}{\sqrt{1 - \frac{v^2}{c^2}}}$$

and v is the velocity of the mass relative to an observer. There are many aspects of the total energy E that we will discuss—among them are how kinetic and potential energies are included in E, and how E is related to relativistic momentum. But first, note that at rest, total energy is not zero. Rather, when $v = 0$, we have $\gamma = 1$, and an object has rest energy.

Rest Energy

Rest energy is $E_0 = mc^2$.

This is the correct form of Einstein's most famous equation, which for the first time showed that energy is related to the mass of an object at rest. For example, if energy is stored in the object, its rest mass increases. This also implies that mass can be destroyed to release energy. The implications of these first two equations regarding relativistic energy are so broad that they were not completely recognized for some years after Einstein published them in 1907, nor was the experimental proof that they are correct widely recognized at first. Einstein, it should be noted, did understand and describe the meanings and implications of his theory.

Kinetic Energy and the Ultimate Speed Limit

Kinetic energy is energy of motion. Classically, kinetic energy has the familiar expression $\frac{1}{2}mv^2$.

The relativistic expression for kinetic energy is obtained from the work-energy theorem. This theorem states that the net work on a system goes into kinetic energy. If our system starts from rest, then the work-energy theorem is W_{net} = KE.

Relativistically, at rest we have rest energy $E_0 = mc^2$. The work increases this to the total energy $E = \gamma mc^2$. Thus, $W_{net} = E - E_0 = \gamma mc^2 - mc^2 = (\gamma - 1)mc^2$.

Relativistically, we have $W_{net} = \text{KE}_{rel}$.

Relativistic Kinetic Energy

Relativistic kinetic energy is $\text{KE}_{rel} = (\gamma - 1)mc^2$.

When motionless, we have $v = 0$ and

$$\gamma = \frac{1}{\sqrt{1-\frac{v^2}{c^2}}} = 1$$

so that $\text{KE}_{rel} = 0$ at rest, as expected. But the expression for relativistic kinetic energy (such as total energy and rest energy) does not look much like the classical $\frac{1}{2}mv^2$. To show that the classical expression for kinetic energy is obtained at low velocities, we note that the binomial expansion for γ at low velocities gives,

$$\gamma = 1 + \frac{1v^2}{2c^2}$$

A binomial expansion is a way of expressing an algebraic quantity as a sum of an infinite series of terms. In some cases, as in the limit of small velocity here, most terms are very small. Thus the expression derived for γ here is not exact, but it is a very accurate approximation. Thus, at low velocities,

$$\gamma - 1 = \frac{1v^2}{2c^2}$$

Entering this into the expression for relativistic kinetic energy gives,

$$\text{KE}_{rel} = \left(\frac{1v^2}{2c^2}\right)mc^2 = \frac{1}{2}mv^2 = \text{KE}_{class}$$

So, in fact, relativistic kinetic energy does become the same as classical kinetic energy when $v << c$.

It is even more interesting to investigate what happens to kinetic energy when the velocity of an object approaches the speed of light. We know that γ becomes infinite as v approaches c, so that KE_{rel} also becomes infinite as the velocity approaches the speed of light. An infinite amount of work (and, hence, an infinite amount of energy input) is required to accelerate a mass to the speed of light.

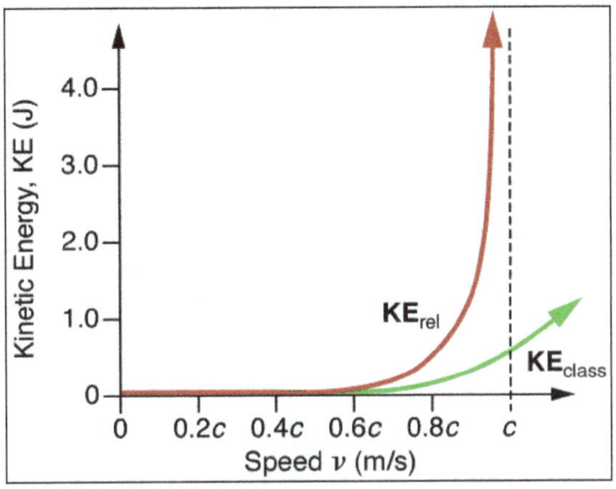

This graph of KE_{rel} versus velocity shows how kinetic energy approaches infinity as velocity approaches the speed of light. It is thus not possible for an object having mass to reach the speed of light. Also shown is KE_{class}, the classical kinetic energy, which is similar to relativistic kinetic energy at low velocities. Note that much more energy is required to reach high velocities than predicted classically.

So the speed of light is the ultimate speed limit for any particle having mass. All of this is consistent with the fact that velocities less than c always add to less than c. Both the relativistic form for kinetic energy and the ultimate speed limit being c have been confirmed in detail in numerous experiments. No matter how much energy is put into accelerating a mass, its velocity can only approach—not reach—the speed of light.

Example

Comparing Kinetic Energy: Relativistic Energy Versus Classical Kinetic Energy

An electron has a velocity $v = 0.990c$.

1. Calculate the kinetic energy in MeV of the electron.
2. Compare this with the classical value for kinetic energy at this velocity. (The mass of an electron is 9.11×10^{-31} kg.)

Strategy

The expression for relativistic kinetic energy is always correct, but for Part 1 it must be used since the velocity is highly relativistic (close to c). First, we will calculate the relativistic factor γ, and then use it to determine the relativistic kinetic energy. For Part 2, we will calculate the classical kinetic energy (which would be close to the relativistic value if v were less than a few percent of c) and see that it is not the same.

Solution for Part 1:

Identify the knowns: $v = 0.990c$; $m = 9.11 \times 10^{-31}$ kg

Identify the unknown: KE_{rel}

Choose the appropriate equation. $KE_{rel} = (\gamma - 1)mc^2$

Plug the knowns into the equation.First calculate γ. We will carry extra digits because this is an intermediate calculation.

$$\begin{aligned}\gamma &= \frac{1}{\sqrt{1-\frac{v^2}{c^2}}}\\ &= \frac{1}{\sqrt{1-\frac{(0.990c)^2}{c^2}}}\\ &= \frac{1}{\sqrt{1-(0.990)^2}}\\ &= 7.0888\end{aligned}$$

Next, we use this value to calculate the kinetic energy.

$$\begin{aligned}KE_{rel} &= (\gamma - 1)mc^2\\ &= (7.0888-1)(9.11\times10^{-31}\,kg)(3.00\times10^{8}\,m/s)^2\\ &= 4.99\times10^{13}\,J\end{aligned}$$

Convert units.

$$\begin{aligned}KE_{rel} &= (4.99\times10^{13}\,J)\left(\frac{1\,MeV}{1.60\times10^{-13}\,J}\right)\\ &= 3.12MeV\end{aligned}$$

Solution for Part 2:

List the knowns: $v = 0.990c$; $m = 9.11 \times 10^{-31}$ kg

List the unknown: KE_{class}

Choose the appropriate equation.

$$KE_{class} = \frac{1}{2}mv^2$$

Plug the knowns into the equation.

$$KE_{class} = \frac{1}{2}mv^2$$

$$= \frac{1}{2}\left(9.00\times10^{-31}\,\text{kg}\right)(0.990)^2\left(3.00\times10^8 m/s\right)^2$$
$$= 4.02\times10^{-14}\,\text{J}$$

Convert units.

$$\text{KE}_{\text{class}} = 4.02\times10^{-14}\,\text{J}\left(\frac{1\,\text{MeV}}{1.60\times10^{-13}\,\text{J}}\right)$$
$$= 0.251\text{MeV}$$

As might be expected, since the velocity is 99.0% of the speed of light, the classical kinetic energy is significantly off from the correct relativistic value. Note also that the classical value is much smaller than the relativistic value. In fact,

$$\frac{\text{KE}_{\text{rel}}}{\text{KE}_{\text{class}}} = 12.4$$

here. This is some indication of how difficult it is to get a mass moving close to the speed of light. Much more energy is required than predicted classically.

The Fermi National Accelerator Laboratory, near Batavia, Illinois, was a subatomic particle collider that accelerated protons and antiprotons to attain energies up to 1 Tev (a trillion electronvolts). The circular ponds near the rings were built to dissipate waste heat

Some people interpret this extra energy as going into increasing the mass of the system, but, as discussed in Relativistic Momentum, this cannot be verified unambiguously. What is certain is that ever-increasing amounts of energy are needed to get the velocity of a mass a little closer to that of light. An energy of 3 MeV is a very small amount for an electron, and it can be achieved with present-day particle accelerators. SLAC, for example, can accelerate electrons to over 50×10^9 eV = 50, 000 MeV.

Is there any point in getting v a little closer to c than 99.0% or 99.9%? The answer is yes. We learn a great deal by doing this. The energy that goes into a high-velocity mass can be converted to any other form, including into entirely new masses. Most of what we know about the substructure of matter and the collection of exotic short-lived particles in nature has been learned this way.

Particles are accelerated to extremely relativistic energies and made to collide with other particles, producing totally new species of particles. Patterns in the characteristics of these previously unknown particles hint at a basic substructure for all matter.

Relativistic Energy and Momentum

We know classically that kinetic energy and momentum are related to each other, since,

$$KE_{class} = \frac{p^2}{2m} = \frac{(mv)^2}{2m} = \frac{1}{2}mv^2$$

Relativistically, we can obtain a relationship between energy and momentum by algebraically manipulating their definitions. This produces $E^2 = (pc)^2 + (mc^2)^2$, where E is the relativistic total energy and p is the relativistic momentum. This relationship between relativistic energy and relativistic momentum is more complicated than the classical, but we can gain some interesting new insights by examining it. First, total energy is related to momentum and rest mass. At rest, momentum is zero, and the equation gives the total energy to be the rest energy mc^2 (so this equation is consistent with the discussion of rest energy above). However, as the mass is accelerated, its momentum p increases, thus increasing the total energy. At sufficiently high velocities, the rest energy term $(mc^2)^2$becomes negligible compared with the momentum term $(pc)^2$; thus, $E = pc$ at extremely relativistic velocities.

If we consider momentum p to be distinct from mass, we can determine the implications of the equation $E^2 = (pc)^2 + (mc^2)^2$, for a particle that has no mass. If we take m to be zero in this equation, then,

$$E = pc, or\ p = \frac{E}{c}.$$

Massless particles have this momentum. There are several massless particles found in nature, including photons (these are quanta of electromagnetic radiation). Another implication is that a massless particle must travel at speed c and only at speed c. While it is beyond the scope of this text to examine the relationship in the equation $E^2 = (pc)^2 + (mc^2)^2$, in detail, we can see that the relationship has important implications in special relativity.

Origins and Derivation of the Equation

The Energy–momentum relation was first established by Paul Dirac in 1928 under the form $E = \sqrt{(c^2p^2 + (m_0c^2)^2)} + V$, where V is the amount of potential energy.

The equation can be derived in a number of ways, two of the simplest include:

1. From the relativistic dynamics of a massive particle.
2. By evaluating the norm of the four-momentum of the system. This method applies to both massive and masseless particles, and can be extended to multi-particle systems with relatively little effort.

Heuristic Approach for Massive Particles

For a massive object moving at three-velocity u = (u_x, u_y, u_z) with magnitude |u| = u in the lab frame:

$$E = \gamma_{(\mathbf{u})} m_0 c^2$$

is the total energy of the moving object in the lab frame,

$$\mathbf{p} = \gamma_{(\mathbf{u})} m_0 \mathbf{u}$$

is the three dimensional relativistic momentum of the object in the lab frame with magnitude |p| = p. The relativistic energy E and momentum p include the Lorentz factor defined by:

$$\gamma_{(\mathbf{u})} = \frac{1}{\sqrt{1 - \frac{\mathbf{u} \cdot \mathbf{u}}{c^2}}} = \frac{1}{\sqrt{1 - \left(\frac{u}{c}\right)^2}}$$

Some authors use relativistic mass defined by:

$$m = \gamma_{(\mathbf{u})} m_0$$

although rest mass m_0 has a more fundamental significance, and will be used primarily over relativistic mass m.

Squaring the 3-momentum gives:

$$p^2 = \mathbf{p} \cdot \mathbf{p} = \frac{m_0^2 \mathbf{u} \cdot \mathbf{u}}{1 - \frac{\mathbf{u} \cdot \mathbf{u}}{c^2}} = \frac{m_0^2 u^2}{1 - \left(\frac{u}{c}\right)^2}$$

then solving for u^2 and substituting into the Lorentz factor obtains its alternative form in terms of 3-momentum and mass, rather than 3-velocity:

$$\gamma = \sqrt{1 + \left(\frac{p}{m_0 c}\right)^2}$$

Inserting this form of the Lorentz factor into the energy equation:

$$E = m_0 c^2 \sqrt{1 + \left(\frac{p}{m_0 c}\right)^2}$$

followed by more rearrangement yields (*1*). The elimination of the Lorentz factor also eliminates implicit velocity dependence of the particle in (*1*), as well as any inferences to the “relativistic mass” of a massive particle. This approach is not general as massless particles are not considered. Naively setting $m_0 = 0$ would mean that $E = 0$ and p = 0 and no energy–momentum relation could be derived, which is not correct.

Norm of the Four-momentum

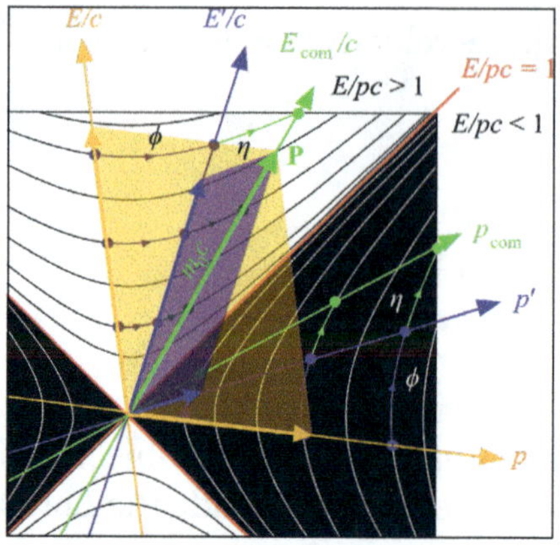

The energy and momentum of an object measured in two inertial frames in energy–momentum space – the yellow frame measures E and p while the blue frame measures E' and p′. The green arrow is the four-momentum P of an object with length proportional to its rest mass m_0. The green frame is the centre-of-momentum frame for the object with energy equal to the rest energy. The hyperbolae show the Lorentz transformation from one frame to another is a hyperbolic rotation, and ϕ and $\phi + \eta$ are the rapidities of the blue and green frames, respectively.

Special Relativity

In Minkowski space, energy (divided by c) and momentum are two components of a Minkowski four-vector, namely the four-momentum;

$$\mathrm{P} = \left(\frac{E}{c}, \mathrm{p}\right),$$

(these are the contravariant components).

The Minkowski inner product $\langle\,,\rangle$ of this vector with itself gives the square of the norm of this vector, it is proportional to the square of the rest mass m of the body:

$$\langle \mathrm{P},\mathrm{P}\rangle = |\mathrm{P}|^2 = (m_0 c)^2,$$

a Lorentz invariant quantity, and therefore independent of the frame of reference. Using the Minkowski metric η with metric signature$(-+++)$, the inner product is,

$$\langle \mathrm{P},\mathrm{P}\rangle = |\mathrm{P}|^2 = -(m_0 c)^2,$$

and

$$\langle \mathrm{P},\mathrm{P}\rangle = P^{\alpha}\eta_{\alpha\beta}P^{\beta} = \begin{pmatrix} \frac{E}{c} & p_x & p_y & p_z \end{pmatrix} \begin{pmatrix} -1 & 0 & 0 & 0 \\ 0 & 1 & 0 & 0 \\ 0 & 0 & 1 & 0 \\ 0 & 0 & 0 & 1 \end{pmatrix} \begin{pmatrix} \frac{E}{c} \\ p_x \\ p_y \\ p_z \end{pmatrix} = -\left(\frac{E}{c}\right)^2 + p^2,$$

so,

$$-(m_o c)^2 = -\left(\frac{E}{c}\right)^2 + p^2.$$

General Relativity

In general relativity, the 4-momentum is a four-vector defined in a local coordinate frame, although by definition the inner product is similar to that of special relativity,

$$\langle \mathrm{P},\mathrm{P}\rangle = |\mathrm{P}|^2 = (m_o c)^2,$$

in which the Minkowski metric η is replaced by the metric tensor field g:

$$\langle \mathrm{P},\mathrm{P}\rangle = |\mathrm{P}|^2 = P^\alpha g_{\alpha\beta} P^\beta,$$

solved from the Einstein field equations. Then:

$$P^\alpha g_{\alpha\beta} P^\beta = (m_o c)^2.$$

Performing the summations over indices followed by collecting "time-like", "spacetime-like", and "space-like" terms gives:

$$\underbrace{g_{oo}\left(P^o\right)^2}_{\text{time-like}} + 2\underbrace{g_{oi} P^o P^i}_{\text{spacetime-like}} + \underbrace{g_{ij} P^i P^j}_{\text{space-like}} = (m_o c)^2.$$

where the factor of 2 arises because the metric is a symmetric tensor, and the convention of Latin indices i, j taking space-like values 1, 2, 3 is used. As each component of the metric has space and time dependence in general; this is significantly more complicated than the formula quoted at the beginning.

Galilean Transformation

Galilean transformations, also called Newtonian transformations, set of equations in classical physics that relate the space and time coordinates of two systems moving at a constant velocityrelative to each other. Adequate to describe phenomena at speeds much smaller than the speed of light, Galilean transformations formally express the ideas that space and time are absolute; that length, time, and mass are independent of the relative motion of the observer; and that the speed of light depends upon the relative motion of the observer.

Suppose there are two reference frames (systems) designated by S and S' such that the co-ordinate axes are parallel. In S, we have the co-ordinates $\{x,y,z,t\}$ and in S' we have the co-ordinates $\{x',y',z',t'\}$. S' is moving with respect to S with velocity v (as measured in S) in the x direction. The clocks in both systems were synchronised at time $t = 0$ and they run at the same rate.

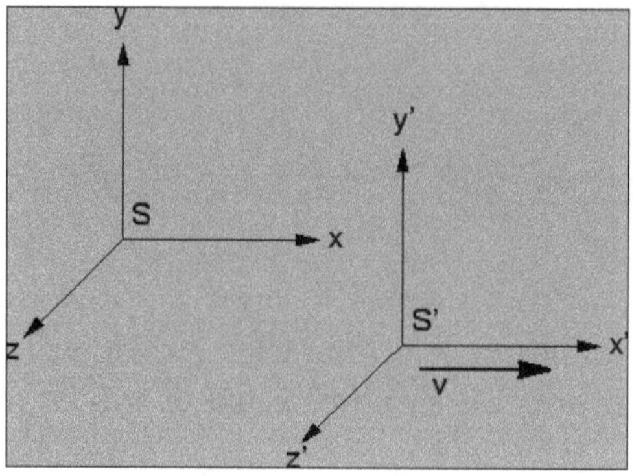

Reference frame S' moves with velocity v (in the x direction) relative to reference frame S.

We have the intuitive relationships,

$$\begin{aligned} x' &= x - vt \\ y' &= y \\ z' &= z \\ t' &= t \end{aligned}$$

This set of equations is known as the Galilean Transformation. They enable us to relate a measurement in one inertial reference frame to another. For example, suppose we measure the velocity of a vehicle moving in the in x-direction in system S, and we want to know what would be the velocity of the vehicle in S'.

$$v'_x = \frac{dx'}{dt'} = \frac{d(x - vt)}{dt} = v_x - v$$

This is the result our intuition is familiar with.

We have stated the we would like the laws of physics to be the same in all inertial reference frames, as this is indeed our experience of nature. Physically, we should be able to perform the same experiments in different reference frames, and find always the same physical laws. Mathematically, these laws are expressed by equations. So, we should be able to "transform" our equations from one inertial reference frame to the other inertial reference frame, and always find the same answer.

Suppose we wanted to check that Newton's Second Law is the same in two different reference frames. We put one observer in the un-primed frame, and the other in the primed frame, moving with velocity v relative to the un-primed frame. Consider the vehicle of the previous case undergoing a constant acceleration in the x-direction,

$$\begin{aligned} f' = m'a' &= m'\frac{d^2x'}{dt'^2} \\ &= m'\frac{d}{dt'}\left(\frac{dx'}{dt'}\right) \\ &= m'\frac{d}{dt}\left(\frac{d(x - vt)}{dt}\right) \end{aligned}$$

$$
\begin{aligned}
&= m\frac{d(v_x - v)}{dt} \\
&= m\frac{dv_x}{dt} \\
&= ma = f
\end{aligned}
$$

Indeed, it does not matter which inertial frame we observe from, we recover the same Second Law of Motion each time. In the parlance of physics, we say the Second Law of Motion is invariant under the Galilean Transformation.

Exercise,

You show that the Law of Momentum Conservation holds regardless of the inertial frame a given collision is viewed in. This is done by specialising to a collision where all velocities are in the x-direction. How would you do this for a more general collision ?

So far so good !

We have Classical Mechanics, a beautiful theory, as it has an elegant independence of how you observe it. There is a sense of poetry in how ugly terms, arising from observation in a different frame, eventually drop away, until we are left with physical laws which are invariant under the Galilean Transformation. But as time passes, it becomes clear we are in a fools paradise.

The first problem. Experiments on electric and magnetic fields, as well as induction of one type of field from changes in the other, lead to the collection of a set of equations, describing all these phenomena, known as Maxwell's Equations. You are already familiar with them. In vacuum they are,

MaxwellsEquations
in vacuo

$$
\begin{aligned}
\nabla.\mathrm{B} &= 0, \\
\nabla.\mathrm{E} &= 0, \\
\nabla\times\mathrm{B} &= \epsilon_0\,\mu_0\frac{\partial E}{\partial t}, \\
\nabla\times\mathrm{E} &= -\frac{\partial \mathrm{B}}{\partial t}.
\end{aligned}
$$

Now, these equations are considered to be rock solid, arising from and verified by many experiments. Amazingly, they imply the existence of a previously not guessed at phenomenon. This is the electromagnetic wave. Every electrical engineer, following Marconi, must appreciate this !

To see this in detail, take the time derivative of the second last equation and the curl of the last.

$$
\begin{aligned}
\frac{\partial}{\partial t}\nabla\times B &= \epsilon_0\,\mu_0\frac{\partial E}{\partial t}, \\
\nabla\times(\nabla\times\mathrm{E}) &= -\nabla\times\frac{\partial B}{\partial t}.
\end{aligned}
$$

Now note that space and time derivatives commute,

$$
\frac{\partial}{\partial t}\nabla\times\mathrm{B} = \nabla\times\frac{\partial B}{\partial t},
$$

so,

$$\nabla\times(\nabla\times\mathrm{E}) = -\epsilon_0\,\mu_0\frac{\partial^2 E}{\partial t^2}.$$

Now, we use the identity,

$$\nabla\times(\nabla\times\mathrm{E}) = \nabla\nabla.\mathrm{E}\text{-}\nabla^2\mathrm{E}.$$

The second term of the above equation drops out due to the vanishing of the divergence of the electric field (the second of Maxwell's Equations). So, we finally have the three dimensional wave equation,

$$\nabla^2\mathrm{E} = \epsilon_0\,\mu_0\frac{\partial^2 E}{\partial t^2}.$$

To see this is a wave equation, note the analogy in one dimension,

$$\frac{\partial^2 y}{\partial x^2} = \frac{1}{c^2}\frac{\partial^2 y}{\partial t^2}.$$

which is solved by the wave function,

$$y(x,t) = \sin(x-ct),$$

which represents a wave traveling along the x axis with velocity c.

It is clear therefore that Maxwell's Equations are highly predictive.

1. A diversity is unified in a simplicity. The various phenomena of radiowaves, microwaves, infrared, visible and ultra-violet light, X-rays and gamma rays are all electromagnetic waves, differing only in their frequency.
2. They all travel at the same speed.
3. Even that speed is specified : $c = 1/\sqrt{\epsilon_0\mu_0} = 2.997\times10^8$ m/s.
4. The speed appears independent of the source and the observer.

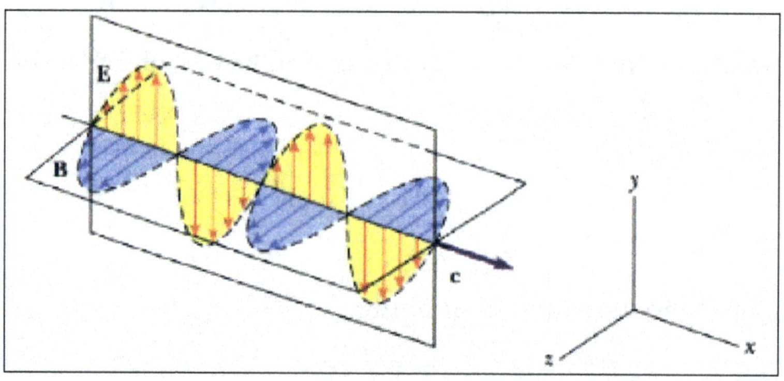

An electromagnetic wave traveling in vacuum.

Lorentz Transformation

Lorentz transformations, set of equations in relativity physics that relate the space and time coordinates of two systems moving at a constant velocity relative to each other. Required to describe high-speed phenomena approaching the speed of light, Lorentz transformations formally express the relativity concepts that space and time are not absolute; that length, time, and mass depend on the relative motion of the observer; and that the speed of light in a vacuum is constant and independent of the motion of the observer or the source. The equations were developed by the Dutch physicist Hendrik Antoon Lorentz in 1904.

These co-ordinate transformations are known as the Lorentz Transformation. Note that the Lorentz Transformation reduces to the Galilean Transformation when $v \ll c$ so that $k = 1$ as required. Lorentz derived this transformation which made the Maxwell Equations of electromagnetism the same in all inertial reference frames in 1895. It was Einstein who is credited (in 1905) with discovering the full significance of these equations : The equations of the Lorentz Transformation define the relativity of space-time when measurements are made in different inertial reference frames.

The fact the the speed of light is "c" (constant) in all inertial reference frames implies a relativity in space-time. We expect length and time measurements to differ in different frames, in such a way that the speed of light is constant.

Because the relativistic effects become more obvious at velocities approaching the speed of light, a special symbol, "*beta*" defined as,

$$\beta = v / c$$

is used in relativistic parlance. Clearly, for velocities such that $t_2 = t_1$, relativistic effects are negligible, and for velocities such that $L = L_0\sqrt{1 - v^2 / c^2} = L_0 / \gamma$, relativistic effects must be considered. The scale of relativity between reference fames is also an important characteristic. This is given the the special symbol "*gamma*" defined as,

$$\gamma = \frac{1}{\sqrt{1 - v^2 / c^2}}$$

Clearly, non-relativistic velocities $t_2 = t_1$ imply $\gamma \approx 1$, whereas extreme relativistic velocities imply $\gamma \gg 1$, as in figure below.

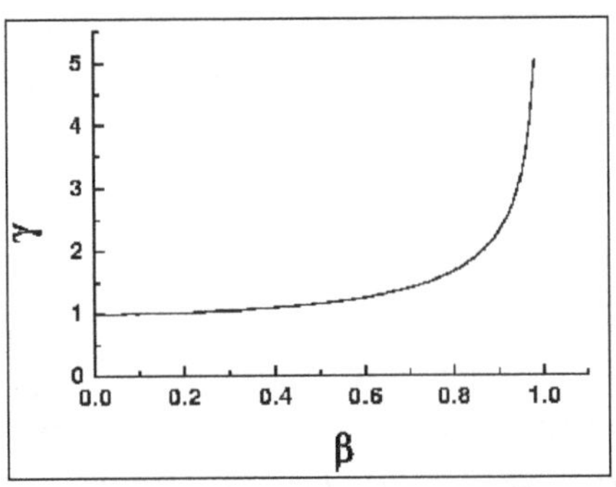

The gradual development of relativistic effects as $v \to c$.

Derivation of the Group of Lorentz Transformations

An *event* is something that happens at a certain point in spacetime, or more generally, the point in spacetime itself. In any inertial frame an event is specified by a time coordinate *ct*and a set of Cartesian coordinates *x*, *y*, *z* to specify position in space in that frame. Subscripts label individual events.

From Einstein's second postulate of relativity follows,

$$c^2(t_2-t_1)^2-(x_2-x_1)^2-(y_2-y_1)^2-(z_2-z_1)^2=0 \quad \text{(lightlike separated events 1, 2)}$$

in all inertial frames for events connected by *light signals*. The quantity on the left is called the *spacetime interval* between events $a_1 = (t_1, x_1, y_1, z_1)$ and $a_2 = (t_2, x_2, y_2, z_2)$. The interval between *any two* events, not necessarily separated by light signals, is in fact invariant, i.e., independent of the state of relative motion of observers in different inertial frames, as is shown using homogeneity and isotropy of space. The transformation sought after thus must possess the property that,

$$\begin{aligned} &c^2(t_2-t_1)^2-(x_2-x_1)^2-(y_2-y_1)^2-(z_2-z_1)^2 \\ &=c^2(t_{2'}-t_{1'})^2-(x_{2'}-x_{1'})^2-(y_{2'}-y_{1'})^2-(z_{2'}-z_{1'})^2 \quad \text{(all events 1, 2).} \end{aligned}$$

where (ct, x, y, z) are the spacetime coordinates used to define events in one frame, and (ct', x', y', z') are the coordinates in another frame. First one observes that equation above is satisfied if an arbitrary 4-tuple b of numbers are added to events a_1 and a_2. Such transformations are called *spacetime translations* and are not dealt with further here. Then one observes that a *linear* solution preserving the origin of the simpler problem,

$$\begin{aligned} &c^2t^2-x^2-y^2-z^2=c^2t'^2-x'^2-y'^2-z'^2 \\ \text{or} \quad &c^2t_1t_2-x_1x_2-y_1y_2-z_1z_2=c^2t_1't_2'-x_1'x_2'-y_1'y_2'-z_1'z_2' \end{aligned}$$

solves the general problem too. Finding the solution to the simpler problem is just a matter of look-up in the theory of classical groups that preserve bilinear forms of various signature.Equation above can be written more compactly as,

$$(a,a)=(a',a') \quad \text{or} \quad a\cdot a=a'\cdot a',$$

where $(\cdot, \cdot)$ refers to the bilinear form of signature (1, 3) on R^4 exposed by the right hand side formula in third set of equations. The alternative notation defined on the right is referred to as the *relativistic dot product*. Spacetime mathematically viewed as R^4 endowed with this bilinear form is known as Minkowski space *M*. The Lorentz transformation is thus an element of the group Lorentz group O(1, 3), the Lorentz group or, for those that prefer the other metric signature, O(3, 1) (also called the Lorentz group). One has,

$$(a,a)=(\Lambda a,\Lambda a)=(a',a'), \quad \Lambda\in O(1,3), \quad a,a'\in M,$$

which is precisely preservation of the bilinear form in third set of equations which implies (by linearity of Λ and bilinearity of the form) that second set of equations is satisfied. The elements of the Lorentz group are rotations and *boosts* and mixes thereof. If the spacetime translations are included, then one obtains the *inhomogeneous Lorentz group* or the Poincaré group.

Generalities

The relations between the primed and unprimed spacetime coordinates are the Lorentz transformations, each coordinate in one frame is a linear function of all the coordinates in the other frame, and the inverse functions are the inverse transformation. Depending on how the frames move relative to each other, and how they are oriented in space relative to each other, other parameters that describe direction, speed, and orientation enter the transformation equations.

Transformations describing relative motion with constant (uniform) velocity and without rotation of the space coordinate axes are called *boosts*, and the relative velocity between the frames is the parameter of the transformation. The other basic type of Lorentz transformations is rotations in the spatial coordinates only, these are also inertial frames since there is no relative motion, the frames are simply tilted (and not continuously rotating), and in this case quantities defining the rotation are the parameters of the transformation (e.g., axis–angle representation, or Euler angles, etc.). A combination of a rotation and boost is a *homogeneous transformation*, which transforms the origin back to the origin.

The full Lorentz group O(3, 1) also contains special transformations that are neither rotations nor boosts, but rather reflections in a plane through the origin. Two of these can be singled out; spatial inversion in which the spatial coordinates of all events are reversed in sign and temporal inversion in which the time coordinate for each event gets its sign reversed.

Boosts should not be conflated with mere displacements in spacetime; in this case, the coordinate systems are simply shifted and there is no relative motion. However, these also count as symmetries forced by special relativity since they leave the spacetime interval invariant. A combination of a rotation with a boost, followed by a shift in spacetime, is an *inhomogeneous Lorentz transformation*, an element of the Poincaré group, which is also called the inhomogeneous Lorentz group.

Physical Formulation of Lorentz Boosts

Coordinate Transformation

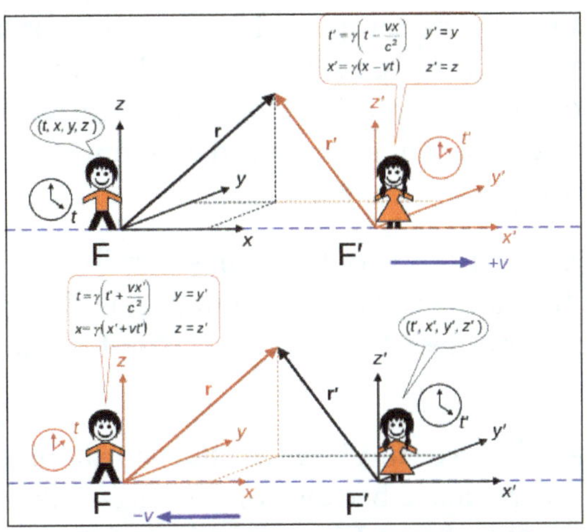

The spacetime coordinates of an event, as measured by each observer in their inertial reference frame (in standard configuration) are shown in the speech bubbles.
Top: frame F' moves at velocity v along the x-axis of frame F.
Bottom: frame F moves at velocity $-v$ along the x'-axis of frame F'.

A "stationary" observer in frame F defines events with coordinates t, x, y, z. Another frame F' moves with velocity v relative to F, and an observer in this "moving" frame F' defines events using the coordinates t', x', y', z'.

The coordinate axes in each frame are parallel (the x and x' axes are parallel, the y and y' axes are parallel, and the z and z'axes are parallel), remain mutually perpendicular, and relative motion is along the coincident xx' axes. At $t = t' = 0$, the origins of both coordinate systems are the same, $(x, y, z) = (x', y', z') = (0, 0, 0)$. In other words, the times and positions are coincident at this event. If all these hold, then the coordinate systems are said to be in standard configuration, or synchronized.

If an observer in F records an event t, x, y, z, then an observer in F' records the *same* event with coordinates.

Lorentz boost (x *direction*):

$$t' = \gamma\left(t - \frac{vx}{c^2}\right)$$
$$x' = \gamma(x - vt)$$
$$y' = y$$
$$z' = z$$

where v is the relative velocity between frames in the x-direction, c is the speed of light, and

$$\gamma = \frac{1}{\sqrt{1 - \frac{v^2}{c^2}}}$$

(lowercase gamma) is the Lorentz factor.

Here, v is the *parameter* of the transformation, for a given boost it is a constant number, but can take a continuous range of values. In the setup used here, positive relative velocity $v > 0$ is motion along the positive directions of the xx' axes, zero relative velocity $v = 0$ is no relative motion, while negative relative velocity $v < 0$ is relative motion along the negative directions of the xx' axes. The magnitude of relative velocity v cannot equal or exceed c, so only subluminal speeds $-c < v < c$ are allowed. The corresponding range of γ is $1 \leq \gamma < \infty$.

The transformations are not defined if v is outside these limits. At the speed of light ($v = c$) γ is infinite, and faster than light ($v > c$) γ is a complex number, each of which make the transformations unphysical. The space and time coordinates are measurable quantities and numerically must be real numbers.

As an active transformation, an observer in F′ notices the coordinates of the event to be "boosted" in the negative directions of the xx' axes, because of the $-v$ in the transformations. This has the equivalent effect of the *coordinate system* F′ boosted in the positive directions of the xx' axes, while the event does not change and is simply represented in another coordinate system, a passive transformation.

The inverse relations (t, x, y, z in terms of t', x', y', z') can be found by algebraically solving the original set of equations. A more efficient way is to use physical principles. Here F' is the "stationary" frame while F is the "moving" frame. According to the principle of relativity, there is no privileged frame of reference, so the transformations from F' to F must take exactly the same form as the transformations from F to F'. The only difference is F moves with velocity $-v$ relative to F' (i.e., the relative velocity has the same magnitude but is oppositely directed). Thus if an observer in F' notes an event t', x', y', z', then an observer in F notes the *same* event with coordinates.

Inverse Lorentz boost (x direction):

$$t=\gamma\left(t'+\frac{vx'}{c^2}\right)$$
$$x=\gamma\left(x'+vt'\right)$$
$$y=y'$$
$$z=z',$$

and the value of γ remains unchanged. This "trick" of simply reversing the direction of relative velocity while preserving its magnitude, and exchanging primed and unprimed variables, always applies to finding the inverse transformation of every boost in any direction.

Sometimes it is more convenient to use $\beta = v/c$ (lowercase beta) instead of v, so that,

$$ct'=\gamma\left(ct-\beta x\right),$$
$$x'=\gamma\left(x-\beta ct\right),$$

which shows much more clearly the symmetry in the transformation. From the allowed ranges of v and the definition of β, it follows $-1 < \beta < 1$. The use of β and γ is standard throughout the literature.

The Lorentz transformations can also be derived in a way that resembles circular rotations in 3d space using the hyperbolic functions. For the boost in the x direction, the results are,

Lorentz boost (*x direction with rapidity* ζ):

$$ct'=ct\cosh\zeta-x\sinh\zeta$$
$$x'=x\cosh\zeta-ct\sinh\zeta$$
$$y'=y$$
$$z'=z$$

where ζ (lowercase zeta) is a parameter called *rapidity* (many other symbols are used, including θ, ϕ, φ, η, ψ, ξ). Given the strong resemblance to rotations of spatial coordinates in 3d space in the Cartesian xy, yz, and zx planes, a Lorentz boost can be thought of as a hyperbolic rotation of spacetime coordinates in the xt, yt, and zt Cartesian-time planes of 4d Minkowski space. The parameter ζ is the hyperbolic angle of rotation, analogous to the ordinary angle for circular rotations. This transformation can be illustrated with a Minkowski diagram.

The hyperbolic functions arise from the *difference* between the squares of the time and spatial coordinates in the spacetime interval, rather than a sum. The geometric significance of the hyperbolic functions can be visualized by taking $x = 0$ or $ct = 0$ in the transformations. Squaring and subtracting the results, one can derive hyperbolic curves of constant coordinate values but varying ζ, which parametrizes the curves according to the identity,

$$cosh^2\zeta - sinh^2\zeta = 1.$$

Conversely the ct and x axes can be constructed for varying coordinates but constant ζ. The definition,

$$tanh\zeta = \frac{sinh\zeta}{cosh\zeta},$$

provides the link between a constant value of rapidity, and the slope of the ct axis in spacetime. A consequence these two hyperbolic formulae is an identity that matches the Lorentz factor,

$$cosh\zeta = \frac{1}{\sqrt{1 - tanh^2\zeta}}.$$

Comparing the Lorentz transformations in terms of the relative velocity and rapidity, or using the above formulae, the connections between β, γ, and ζ are,

$$\beta = tanh\zeta,$$
$$\gamma = cosh\zeta,$$
$$\beta\gamma = sinh\zeta.$$

Taking the inverse hyperbolic tangent gives the rapidity,

$$\zeta = tanh^{-1}\beta.$$

Since $-1 < \beta < 1$, it follows $-\infty < \zeta < \infty$. From the relation between ζ and β, positive rapidity $\zeta > 0$ is motion along the positive directions of the xx' axes, zero rapidity $\zeta = 0$ is no relative motion, while negative rapidity $\zeta < 0$ is relative motion along the negative directions of the xx' axes.

The inverse transformations are obtained by exchanging primed and unprimed quantities to switch the coordinate frames, and negating rapidity $\zeta \to -\zeta$ since this is equivalent to negating the relative velocity. Therefore,

Inverse Lorentz boost (*x direction with rapidity* ζ):

$$ct = ct'cosh\zeta + x'sinh\zeta$$
$$x = x'cosh\zeta + ct'sinh\zeta$$

$$y = y'$$
$$z = z'$$

The inverse transformations can be similarly visualized by considering the cases when $x' = 0$ and $ct' = 0$.

So far the Lorentz transformations have been applied to *one event.* If there are two events, there is a spatial separation and time interval between them. It follows from the linearity of the Lorentz transformations that two values of space and time coordinates can be chosen, the Lorentz transformations can be applied to each, then subtracted to get the Lorentz transformations of the differences;

$$\Delta t' = \gamma\left(\Delta t - \frac{v\Delta x}{c^2}\right),$$

$$\Delta x' = \gamma(\Delta x - v\Delta t),$$

with inverse relations,

$$\Delta t = \gamma\left(\Delta t' + \frac{v\Delta x'}{c^2}\right),$$

$$\Delta x = \gamma(\Delta x' + v\Delta t').$$

where Δ (uppercase delta) indicates a difference of quantities; e.g., $\Delta x = x_2 - x_1$ for two values of x coordinates, and so on.

These transformations on *differences* rather than spatial points or instants of time are useful for a number of reasons:

- In calculations and experiments, it is lengths between two points or time intervals that are measured or of interest (e.g., the length of a moving vehicle, or time duration it takes to travel from one place to another),
- The transformations of velocity can be readily derived by making the difference infinitesimally small and dividing the equations, and the process repeated for the transformation of acceleration,
- If the coordinate systems are never coincident (i.e., not in standard configuration), and if both observers can agree on an event t_0, x_0, y_0, z_0 in F and t_0', x_0', y_0', z_0' in F', then they can use that event as the origin, and the spacetime coordinate differences are the differences between their coordinates and this origin, e.g., $\Delta x = x - x_0$, $\Delta x' = x' - x_0'$, etc.

Physical Implications

A critical requirement of the Lorentz transformations is the invariance of the speed of light, a fact used in their derivation, and contained in the transformations themselves. If in F the equation for a pulse of light along the x direction is $x = ct$, then in F' the Lorentz transformations give $x' = ct'$, and vice versa, for any $-c < v < c$.

For relative speeds much less than the speed of light, the Lorentz transformations reduce to the Galilean transformation.

$$t' \approx t$$

$$x' \approx x - vt$$

in accordance with the correspondence principle. It is sometimes said that nonrelativistic physics is a physics of "instantaneous action at a distance".

Three unintuitive, but correct, predictions of the transformations are mentioned below.

Time Dilation

Suppose there is a clock at rest in F. If a time interval (say a "tick") is measured at the same point so that $\Delta x = 0$, then the transformations give this tick in F' by $\Delta t' = \gamma\Delta t$. Conversely, suppose there is a clock at rest in F'. If a tick is measured at the same point so that $\Delta x' = 0$, then the transformations give this tick in F by $\Delta t = \gamma\Delta t'$. Either way, the boosted observer measures longer time intervals than the observer in the other frame.

Relativity of Simultaneity

Suppose two events occur simultaneously ($\Delta t = 0$) along the x axis, but separated by a nonzero displacement Δx. Then in F', we find that $\Delta t' = \gamma\frac{-v\Delta x}{c^2}$, so the events are no longer simultaneous according to a moving observer.

Length Contraction

Suppose there is a rod at rest in F aligned along the x axis, with length Δx. In F', the rod moves with velocity $-v$, so its length must be measured by taking two simultaneous ($\Delta t' = 0$) measurements at opposite ends. Under these conditions, the inverse Lorentz transform shows that $\Delta x = \gamma\Delta x'$. In F the two measurements are no longer simultaneous, but this does not matter because the rod is at rest in F. We conclude that the boosted observer measures a shorter length, by a factor of γ, than the observer in the rest frame of the rod. Length contraction affects any geometric quantity related to lengths, so from the perspective of a moving observer, areas and volumes will also appear to shrink along the direction of motion.

Vector Transformations

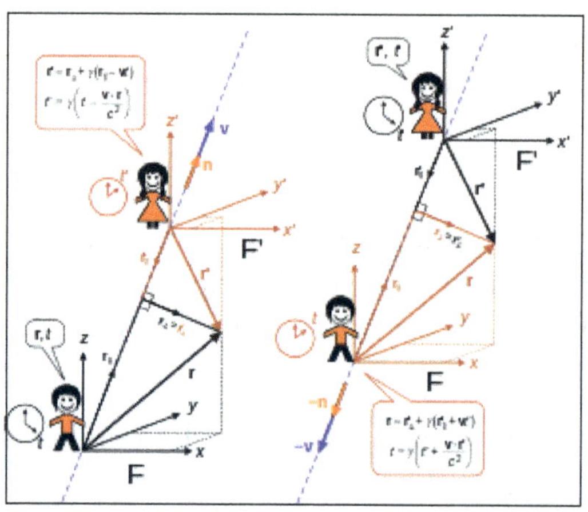

An observer in frame F observes F' to move with velocity v, while F' observes F to move with velocity −v. The coordinate axes of each frame are still parallel and orthogonal. The position vector as measured in each frame is split into components parallel and perpendicular to the relative velocity vector v. Left: Standard configuration. Right: Inverse configuration.

The use of vectors allows positions and velocities to be expressed in arbitrary directions compactly. A single boost in any direction depends on the full relative velocity vector v with a magnitude |v| = v that cannot equal or exceed c, so that $0 \le v < c$.

Only time and the coordinates parallel to the direction of relative motion change, while those coordinates perpendicular do not. With this in mind, split the spatial position vector r as measured in F, and r′ as measured in F', each into components perpendicular (⊥) and parallel (||) to v,

$$\mathrm{r} = \mathrm{r}_\perp + \mathrm{r}_\parallel, \quad \mathrm{r}' = \mathrm{r}_{\perp'} + \mathrm{r}_{\parallel'},$$

then the transformations are,

$$t' = \gamma\left(t - \frac{\mathrm{r}_\parallel \cdot \mathrm{v}}{c^2}\right)$$
$$\mathrm{r}_{\parallel'} = \gamma(\mathrm{r}_\parallel - \mathrm{v}t)$$
$$\mathrm{r}_{\perp'} = \mathrm{r}_\perp$$

where · is the dot product. The Lorentz factor γ retains its definition for a boost in any direction, since it depends only on the magnitude of the relative velocity. The definition $\beta = \mathrm{v}/c$ with magnitude $0 \le \beta < 1$ is also used by some authors.

Introducing a unit vector $\mathrm{n} = \mathrm{v}/v = \beta/\beta$ in the direction of relative motion, the relative velocity is $\mathrm{v} = v\mathrm{n}$ with magnitude v and direction n, and vector projection and rejection give respectively,

$$\mathrm{r}_\parallel = (\mathrm{r}\cdot\mathrm{n})\mathrm{n}, \quad \mathrm{r}_\perp = \mathrm{r} - (\mathrm{r}\cdot\mathrm{n})\mathrm{n}$$

Accumulating the results gives the full transformations,

Lorentz boost (*in direction* n *with magnitude* v):

$$t' = \gamma\left(t - \frac{v\mathrm{n}\cdot\mathrm{r}}{c^2}\right),$$
$$\mathrm{r}' = \mathrm{r} + (\gamma - 1)(\mathrm{r}\cdot\mathrm{n})\mathrm{n} - \gamma t v\mathrm{n}.$$

The projection and rejection also applies to r′. For the inverse transformations, exchange r and r′ to switch observed coordinates, and negate the relative velocity $\mathrm{v} \to -\mathrm{v}$ (or simply the unit vector $\mathrm{n} \to -\mathrm{n}$ since the magnitude v is always positive) to obtain,

Inverse Lorentz boost (*in direction* n *with magnitude* v):

$$t = \gamma\left(t' + \frac{\mathrm{r}'\cdot v\mathrm{n}}{c^2}\right),$$
$$\mathrm{r} = \mathrm{r}' + (\gamma - 1)(\mathrm{r}'\cdot\mathrm{n})\mathrm{n} + \gamma t' v\mathrm{n},$$

The unit vector has the advantage of simplifying equations for a single boost, allows either v or β to be reinstated when convenient, and the rapidity parametrization is immediately obtained by replacing β and $\beta\gamma$. It is not convenient for multiple boosts.

The vectorial relation between relative velocity and rapidity is,

$$\hat{a} = \beta \mathrm{n} = \mathrm{n}\,tanh\,\zeta,$$

and the "rapidity vector" can be defined as,

$$æ = \zeta \mathrm{n} = \mathrm{n}\,tanh^{-1}\beta,$$

each of which serves as a useful abbreviation in some contexts. The magnitude of ζ is the absolute value of the rapidity scalar confined to $0 \le \zeta < \infty$, which agrees with the range $0 \le \beta < 1$.

Transformation of Velocities

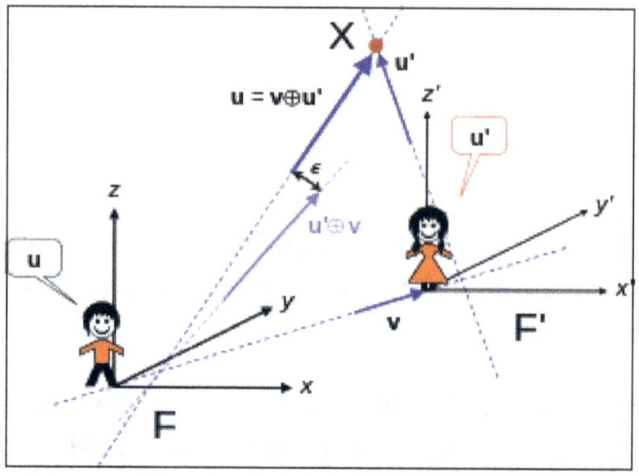

The transformation of velocities provides the definition relativistic velocity addition ⊕, the ordering of vectors is chosen to reflect the ordering of the addition of velocities; first v (the velocity of F′ relative to F) then **u′** (the velocity of X relative to F′) to obtain u = v ⊕ u′ (the velocity of X relative to F).

Defining the coordinate velocities and Lorentz factor by,

$$\mathrm{u} = \frac{dr}{dt}, \quad \mathrm{u}' = \frac{dr'}{dt'}, \quad \gamma_{\mathrm{v}} = \frac{1}{\sqrt{1 - \frac{\mathrm{v}\cdot\mathrm{v}}{c^2}}}$$

taking the differentials in the coordinates and time of the vector transformations, then dividing equations, leads to,

$$\mathrm{u}' = \frac{1}{1 - \frac{\mathrm{v}\cdot\mathrm{u}}{c^2}}\left[\frac{\mathrm{u}}{\gamma_{\mathrm{v}}} - \mathrm{v} + \frac{1}{c^2}\frac{\gamma_{\mathrm{v}}}{\gamma_{\mathrm{v}}+1}(\mathrm{u}\cdot\mathrm{v})\mathrm{v}\right]$$

The velocities u and u′ are the velocity of some massive object. They can also be for a third inertial frame (say *F″*), in which case they must be *constant*. Denote either entity by X. Then X moves with velocity u relative to F, or equivalently with velocity u′relative to F′, in turn F′ moves with velocity v relative to F. The inverse transformations can be obtained in a similar way, or as with position coordinates exchange u and u′, and change B to −v.

The transformation of velocity is useful in stellar aberration, the Fizeau experiment, and the relativistic Doppler effect.

The Lorentz transformations of acceleration can be similarly obtained by taking differentials in the velocity vectors, and dividing these by the time differential.

Transformation of other Quantities

In general, given four quantities A and $\mathbf{Z} = (Z_x, Z_y, Z_z)$ and their Lorentz-boosted counterparts A' and $\mathbf{Z}' = (Z'_x, Z'_y, Z'_z)$, a relation of the form,

$$A^2 - \mathbf{Z}\cdot\mathbf{Z} = A'^2 - \mathbf{Z}'\cdot\mathbf{Z}'$$

implies the quantities transform under Lorentz transformations similar to the transformation of spacetime coordinates;

$$A' = \gamma\left(A - \frac{v\mathbf{n}\cdot\mathbf{Z}}{c}\right),$$

$$\mathbf{Z}' = \mathbf{Z} + (\gamma - 1)(\mathbf{Z}\cdot\mathbf{n})\mathbf{n} - \frac{\gamma A v\mathbf{n}}{c}.$$

The decomposition of Z (and Z′) into components perpendicular and parallel to v is exactly the same as for the position vector, as is the process of obtaining the inverse transformations (exchange (A, Z) and (A', Z′) to switch observed quantities, and reverse the direction of relative motion by the substitution $\mathbf{n} \mapsto -\mathbf{n}$).

The quantities (A, Z) collectively make up a *four vector*, where A is the "timelike component", and Z the "spacelike component". Examples of A and Z are the following:

Four vector	A	Z
Position four vector	Time (multiplied by c), ct	Position vector, r
Four momentum	Energy (divided by c), E/c	Momentum, p
Four-wave vector	angular frequency (divided by c), ω/c	wave vector, k
Four spin	(No name), s_t	Spin, s
Four current	Charge density (multiplied by c), ρc	Current density, j
Electromagnetic four potential	Electric potential (divided by c), φ/c	Magnetic potential, A

For a given object (e.g., particle, fluid, field, material), if A or Z correspond to properties specific to the object like its charge density, mass density, spin, etc., its properties can be fixed in the rest frame of that object. Then the Lorentz transformations give the corresponding properties in a frame moving relative to the object with constant velocity. This breaks some notions taken for granted in non-relativistic physics. For example, the energy E of an object is a scalar in non-relativistic mechanics, but not in relativistic mechanics because energy changes under Lorentz transformations; its value is different for various inertial frames. In the rest frame of an object, it has a rest energy and zero momentum. In a boosted frame its energy is different and it appears to have a momentum. Similarly, in non-relativistic quantum mechanics the spin of a particle is a constant vector, but in relativistic quantum mechanics spin s depends on relative motion. In the rest frame of the particle, the spin pseudovector can be fixed to be its ordinary non-relativistic spin with a zero timelike quantity s_t, however a boosted observer will perceive a nonzero timelike component and an altered spin.

Not all quantities are invariant in the form as shown above, for example orbital angular momentum L does not have a timelike quantity, and neither does the electric field E nor the magnetic field B. The definition of angular momentum is L = r × p, and in a boosted frame the altered angular momentum is L′ = r′ × p′. Applying this definition using the transformations of coordinates and momentum leads to the transformation of angular momentum. It turns out L transforms with another vector quantity N = (E/c^2)r − tp related to boosts. For the case of the E and B fields, the transformations cannot be obtained as directly using vector algebra. The Lorentz force is the definition of these fields, and in F it is F = q(E + v × B) while in F' it is F′ = q(E′ + v′ × B′).

Tensor Formulation

Contravariant Vectors

Writing the general matrix transformation of coordinates as the matrix equation,

$$\begin{bmatrix} x'^0 \\ x'^1 \\ x'^2 \\ x'^3 \end{bmatrix} = \begin{bmatrix} \Lambda^0{}_0 & \Lambda^0{}_1 & \Lambda^0{}_2 & \Lambda^0{}_3 \\ \Lambda^1{}_0 & \Lambda^1{}_1 & \Lambda^1{}_2 & \Lambda^1{}_3 \\ \Lambda^2{}_0 & \Lambda^2{}_1 & \Lambda^2{}_2 & \Lambda^2{}_3 \\ \Lambda^3{}_0 & \Lambda^3{}_1 & \Lambda^3{}_2 & \Lambda^3{}_3 \end{bmatrix} \begin{bmatrix} x^0 \\ x^1 \\ x^2 \\ x^3 \end{bmatrix}$$

allows the transformation of other physical quantities that cannot be expressed as four-vectors; e.g., tensors or spinors of any order in 4d spacetime, to be defined. In the corresponding tensor index notation, the above matrix expression is,

$$x'^{\nu} = \Lambda^{\nu}{}_{\mu} x^{\mu},$$

where lower and upper indices label covariant and contravariant components respectively, and the summation convention is applied. It is a standard convention to use Greekindices that take the value 0 for time components, and 1, 2, 3 for space components, while Latin indices simply take the values 1, 2, 3, for spatial components. Note that the first index (reading left to right) corresponds in the matrix notation to a *row index*. The second index corresponds to the column index.

The transformation matrix is universal for all four-vectors, not just 4-dimensional spacetime coordinates. If A is any four-vector, then in tensor index notation,

$$A'^{\nu} = \Lambda^{\nu}{}_{\mu} A^{\mu}.$$

Alternatively, one writes,

$$A^{\nu'} = \Lambda^{\nu'}{}_{\mu} A^{\mu}.$$

in which the primed indices denote the indices of A in the primed frame. This notation cuts risk of exhausting the Greek alphabet roughly in half.

For a general n-component object one may write,

$$X'^{\alpha} = \Pi(\Lambda)^{\alpha}{}_{\beta} X^{\beta},$$

where Π is the appropriate representation of the Lorentz group, an $n \times n$ matrix for every Λ. In this case, the indices should *not* be thought of as spacetime indices (sometimes called Lorentz indices), and they run from 1 to n. E.g., if X is a bispinor, then the indices are called *Dirac indices*.

Covariant Vectors

There are also vector quantities with covariant indices. They are generally obtained from their corresponding objects with contravariant indices by the operation of *lowering an index*; e.g.,

$$x_\nu = \eta_{\mu\nu} x^\mu,$$

where η is the metric tensor. The inverse of this transformation is given by,

$$x^\mu = \eta^{\nu\mu} x_\nu,$$

where, when viewed as matrices, $\eta^{\mu\nu}$ is the inverse of $\eta_{\mu\nu}$. As it happens, $\eta^{\mu\nu} = \eta_{\mu\nu}$. This is referred to as *raising an index*. To transform a covariant vector A_μ, first raise its index, then transform it according to the same rule as for contravariant 4-vectors, then finally lower the index;

$$A'_\nu = \eta_{\rho\nu} \Lambda^\rho{}_\sigma \eta^{\mu\sigma} A_\mu.$$

But

$$\eta_{\rho\nu} \Lambda^\rho{}_\sigma \eta^{\mu\sigma} = \left(\Lambda^{-1}\right)^\mu{}_\nu,$$

i.e., it is the (μ, ν)-component of the *inverse* Lorentz transformation. One defines (as a matter of notation),

$$\Lambda_\nu{}^\mu \equiv \left(\Lambda^{-1}\right)^\mu{}_\nu,$$

and may in this notation write,

$$A'_\nu = \Lambda_\nu{}^\mu A_\mu.$$

Now for a subtlety. The implied summation on the right hand side of,

$$A'_\nu = \Lambda_\nu{}^\mu A_\mu = \left(\Lambda^{-1}\right)^\mu{}_\nu A_\mu$$

is running over *a row index* of the matrix representing Λ^{-1}. Thus, in terms of matrices, this transformation should be thought of as the *inverse transpose* of Λ acting on the column vector A_μ. That is, in pure matrix notation,

$$A' = \left(\Lambda^{-1}\right)^{\mathrm{T}} A.$$

This means exactly that covariant vectors (thought of as column matrices) transform according to the dual representation of the standard representation of the Lorentz group. This notion generalizes to general representations, simply replace Λ with $\Pi(\Lambda)$.

Tensors

If A and B are linear operators on vector spaces U and V, then a linear operator $A \otimes B$ may be defined on the tensor product of U and V, denoted $U \otimes V$ according to,

$$(A \otimes B)(u \otimes v) = Au \otimes Bv, \qquad u \in U, v \in V, u \otimes v \in U \otimes V.$$

From this it is immediately clear that if u and v are a four-vectors in V, then $u \otimes v \in T_2V \equiv V \otimes V$ transforms as,

$$u \otimes v \to \Lambda u \otimes \Lambda v = \Lambda^{\mu}{}_{\nu} u^{\nu} \otimes \Lambda^{\rho}{}_{\sigma} v^{\sigma} = \Lambda^{\mu}{}_{\nu} \Lambda^{\rho}{}_{\sigma} u^{\nu} \otimes v^{\sigma} \equiv \Lambda^{\mu}{}_{\nu} \Lambda^{\rho}{}_{\sigma} w^{\nu\sigma}.$$

The second step uses the bilinearity of the tensor product and the last step defines a 2-tensor on component form, or rather, it just renames the tensor $u \otimes v$.

These observations generalize in an obvious way to more factors, and using the fact that a general tensor on a vector space V can be written as a sum of a coefficient (component!) times tensor products of basis vectors and basis covectors, one arrives at the transformation law for any tensor quantity T. It is given by,

$$T^{\alpha'\beta'\cdots\zeta'}_{\theta'\iota'\cdots\kappa'} = \Lambda^{\alpha'}{}_{\mu} \Lambda^{\beta'}{}_{\nu} \cdots \Lambda^{\zeta'}{}_{\rho} \Lambda_{\theta'}{}^{\sigma} \Lambda_{\iota'}{}^{\upsilon} \cdots \Lambda_{\kappa'}{}^{\zeta} T^{\mu\nu\cdots\rho}_{\sigma\upsilon\cdots\zeta},$$

where $\Lambda_{\chi'}{}^{\psi}$ is defined above. This form can generally be reduced to the form for general n-component objects given above with a single matrix ($\Pi(\Lambda)$) operating on column vectors. This latter form is sometimes preferred; e.g., for the electromagnetic field tensor.

Transformation of the Electromagnetic Field

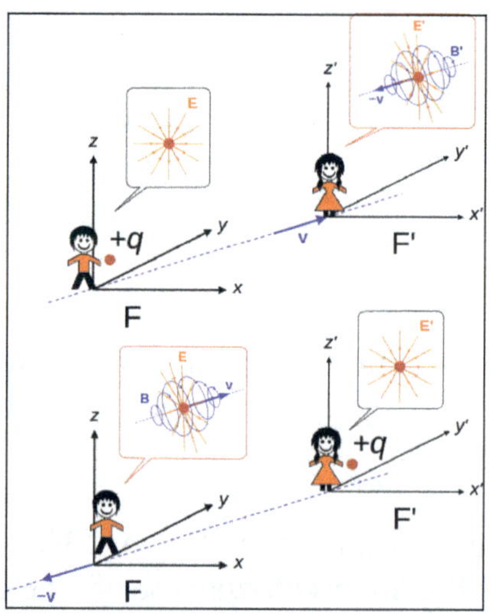

Lorentz boost of an electric charge, the charge is at rest in one frame or the other.

Lorentz transformations can also be used to illustrate that the magnetic field B and electric field E are simply different aspects of the same force — the electromagnetic force, as a consequence of relative motion between electric charges and observers.The fact that the electromagnetic field shows relativistic effects becomes clear by carrying out a simple thought experiment.

- An observer measures a charge at rest in frame F. The observer will detect a static electric field. As the charge is stationary in this frame, there is no electric current, so the observer does not observe any magnetic field.
- The other observer in frame F′ moves at velocity v relative to F and the charge. *This* observer sees a different electric field because the charge moves at velocity −v in their rest frame. The motion of the charge corresponds to an electric current, and thus the observer in frame F′ also sees a magnetic field.

The electric and magnetic fields transform differently from space and time, but exactly the same way as relativistic angular momentum and the boost vector.

The electromagnetic field strength tensor is given by,

$$F^{\mu\nu} = \begin{bmatrix} 0 & -\frac{1}{c}E_x & -\frac{1}{c}E_y & -\frac{1}{c}E_z \\ \frac{1}{c}E_x & 0 & -B_z & B_y \\ \frac{1}{c}E_y & B_z & 0 & -B_x \\ \frac{1}{c}E_z & -B_y & B_x & 0 \end{bmatrix} \text{(SI units, signature } (+,-,-,-)\text{)}.$$

in SI units. In relativity, the Gaussian system of units is often preferred over SI units, even in texts whose main choice of units is SI units, because in it the electric field E and the magnetic induction B have the same units making the appearance of the electromagnetic field tensor more natural. Consider a Lorentz boost in the x-direction. It is given by,

$$\Lambda^{\mu}{}_{\nu} = \begin{bmatrix} \gamma & -\gamma\beta & 0 & 0 \\ -\gamma\beta & \gamma & 0 & 0 \\ 0 & 0 & 1 & 0 \\ 0 & 0 & 0 & 1 \end{bmatrix}, \qquad F^{\mu\nu} = \begin{bmatrix} 0 & E_x & E_y & E_z \\ -E_x & 0 & B_z & -B_y \\ -E_y & -B_z & 0 & B_x \\ -E_z & B_y & -B_x & 0 \end{bmatrix} \text{(Gaussian units, signature } (-,+,+,+)\text{)},$$

where the field tensor is displayed side by side for easiest possible reference in the manipulations below.

The general transformation law $T^{\alpha'\beta'\cdots\zeta'}_{\theta'\iota'\cdots\kappa'} = \Lambda^{\alpha'}{}_{\mu}\Lambda^{\beta'}{}_{\nu}\cdots\Lambda^{\zeta'}{}_{\rho}\Lambda_{\theta'}{}^{\sigma}\Lambda_{\iota'}{}^{\upsilon}\cdots\Lambda_{\kappa'}{}^{\zeta}T^{\mu\nu\cdots\rho}_{\sigma\upsilon\cdots\zeta}$, becomes,

$$F^{\mu'\nu'} = \Lambda^{\mu'}{}_{\mu}\Lambda^{\nu'}{}_{\nu}F^{\mu\nu}.$$

For the magnetic field one obtains,

$$\begin{aligned} B_{x'} &= F^{2'3'} = \Lambda^{2}{}_{\mu}\Lambda^{3}{}_{\nu}F^{\mu\nu} = \Lambda^{2}{}_{2}\Lambda^{3}{}_{3}F^{23} = 1\times1\times B_x \\ &= B_x, \\ B_{y'} &= F^{3'1'} = \Lambda^{3}{}_{\mu}\Lambda^{1}{}_{\nu}F^{\mu\nu} = \Lambda^{3}{}_{3}\Lambda^{1}{}_{\nu}F^{3\nu} = \Lambda^{3}{}_{3}\Lambda^{1}{}_{0}F^{30} + \Lambda^{3}{}_{3}\Lambda^{1}{}_{1}F^{31} \\ &= 1\times(-\beta\gamma)(-E_z) + 1\times\gamma B_y = \gamma B_y + \beta\gamma E_z \\ &= \gamma\left(B - \beta\times E\right)_y \end{aligned}$$

$$\begin{aligned}B_{z'} &= F^{1'2'} = \Lambda^{1}{}_{\mu}\Lambda^{2}{}_{\nu}F^{\mu\nu} = \Lambda^{1}{}_{\mu}\Lambda^{2}{}_{2}F^{\mu 2} = \Lambda^{1}{}_{0}\Lambda^{2}{}_{2}F^{02} + \Lambda^{1}{}_{1}\Lambda^{2}{}_{2}F^{12}\\ &= (-\gamma\beta)\times 1\times E_y + \gamma\times 1\times B_z = \gamma B_z - \beta\gamma E_y\\ &= \gamma(B - \beta\times E)_z\end{aligned}$$

For the electric field results,

$$\begin{aligned}E_{x'} &= F^{0'1'} = \Lambda^{0}{}_{\mu}\Lambda^{1}{}_{\nu}F^{\mu\nu} = \Lambda^{0}{}_{1}\Lambda^{1}{}_{0}F^{10} + \Lambda^{0}{}_{0}\Lambda^{1}{}_{1}F^{01}\\ &= (-\gamma\beta)(-\gamma\beta)(-E_x) + \gamma\gamma E_x = -\gamma^2\beta^2(E_x) + \gamma^2 E_x = E_x(1-\beta^2)\gamma^2\\ &= E_x,\\ E_{y'} &= F^{0'2'} = \Lambda^{0}{}_{\mu}\Lambda^{2}{}_{\nu}F^{\mu\nu} = \Lambda^{0}{}_{\mu}\Lambda^{2}{}_{2}F^{\mu 2} = \Lambda^{0}{}_{0}\Lambda^{2}{}_{2}F^{02} + \Lambda^{0}{}_{1}\Lambda^{2}{}_{2}F^{12}\\ &= \gamma\times 1\times E_y + (-\beta\gamma)\times 1\times B_z = \gamma E_y - \beta\gamma B_z\\ &= \gamma(E + \beta\times B)_y\\ E_{z'} &= F^{0'3'} = \Lambda^{0}{}_{\mu}\Lambda^{3}{}_{\nu}F^{\mu\nu} = \Lambda^{0}{}_{\mu}\Lambda^{3}{}_{3}F^{\mu 3} = \Lambda^{0}{}_{0}\Lambda^{3}{}_{3}F^{03} + \Lambda^{0}{}_{1}\Lambda^{3}{}_{3}F^{13}\\ &= \gamma\times 1\times E_z - \beta\gamma\times 1\times(-B_y) = \gamma E_z + \beta\gamma B_y\\ &= \gamma(E + \beta\times B)_z.\end{aligned}$$

Here, $\beta = (\beta, 0, 0)$ is used. These results can be summarized by,

$$\begin{aligned}E_{\parallel'} &= E_{\parallel}\\ B_{\parallel'} &= B_{\parallel}\\ E_{\perp'} &= \gamma(E_{\perp} + \beta\times B_{\perp}) = \gamma(E + \beta\times B)_{\perp},\\ B_{\perp'} &= \gamma(B_{\perp} - \beta\times E_{\perp}) = \gamma(B - \beta\times E)_{\perp},\end{aligned}$$

and are independent of the metric signature. For SI units, substitute $E \rightarrow E/c$. Misner, Thorne & Wheeler (1973) refer to this last form as the 3 + 1 view as opposed to the *geometric view* represented by the tensor expression,

$$F^{\mu'\nu'} = \Lambda^{\mu'}{}_{\mu}\Lambda^{\nu'}{}_{\nu}F^{\mu\nu},$$

and make a strong point of the ease with which results that are difficult to achieve using the 3 + 1 view can be obtained and understood. Only objects that have well defined Lorentz transformation properties (in fact under *any* smooth coordinate transformation) are geometric objects. In the geometric view, the electromagnetic field is a six-dimensional geometric object in *spacetime* as opposed to two interdependent, but separate, 3-vector fields in *space* and *time*. The fields E (alone) and B (alone) do not have well defined Lorentz transformation properties. The mathematical underpinnings are two equations mentioned in the starting of the topic that immediately yield the third equation mentioned in the starting of the topic. One should note that the primed and unprimed tensors refer to the *same event in spacetime*. Thus the complete equation with spacetime dependence is,

$$F^{\mu'\nu'}(x') = \Lambda^{\mu'}{}_{\mu}\Lambda^{\nu'}{}_{\nu}F^{\mu\nu}\left(\Lambda^{-1}x'\right) = \Lambda^{\mu'}{}_{\mu}\Lambda^{\nu'}{}_{\nu}F^{\mu\nu}(x).$$

Length contraction has an effect on charge density ρ and current density J, and time dilation has an effect on the rate of flow of charge (current), so charge and current distributions must transform in a related way under a boost. It turns out they transform exactly like the space-time and energy-momentum four-vectors,

$$\mathbf{j}' = \mathbf{j} - \gamma\rho v\mathbf{n} + (\gamma - 1)(\mathbf{j}\cdot\mathbf{n})\mathbf{n}$$
$$\rho' = \gamma\left(\rho - \mathbf{j}\cdot\frac{v\mathbf{n}}{c^2}\right),$$

or, in the simpler geometric view,

$$j^{\mu'} = \Lambda^{\mu'}{}_{\mu} j^{\mu}.$$

One says that charge density transforms as the time component of a four-vector. It is a rotational scalar. The current density is a 3-vector.

The Maxwell equations are invariant under Lorentz transformations.

Spinors

Equation $(A \otimes B)(u \otimes v) = Au \otimes Bv, \quad u \in U, v \in V, u \otimes v \in U \otimes V.$ hold unmodified for any representation of the Lorentz group, including the bispinor representation. In $u \otimes v \to \Lambda u \otimes \Lambda v = \Lambda^{\mu}{}_{\nu} u^{\nu} \otimes \Lambda^{\rho}{}_{\sigma} v^{\sigma} = \Lambda^{\mu}{}_{\nu} \Lambda^{\rho}{}_{\sigma} u^{\nu} \otimes v^{\sigma} \equiv \Lambda^{\mu}{}_{\nu} \Lambda^{\rho}{}_{\sigma} w^{\nu\sigma}$. one simply replaces all occurrences of Λ by the bispinor representation $\Pi(\Lambda)$,

$$u \otimes v \to \Pi(\Lambda) u \otimes \Pi(\Lambda) v = \Pi(\Lambda)^{\alpha}{}_{\beta} u^{\beta} \otimes \Pi(\Lambda)^{\rho}{}_{\sigma} v^{\sigma} = \Pi(\Lambda)^{\alpha}{}_{\beta} \Pi(\Lambda)^{\rho}{}_{\sigma} u^{\beta} \otimes v^{\sigma} \equiv \Pi(\Lambda)^{\alpha}{}_{\beta} \Pi(\Lambda)^{\rho}{}_{\sigma} w^{\alpha\beta}.$$

The above equation could, for instance, be the transformation of a state in Fock space describing two free electrons.

Transformation of General Fields

A general *noninteracting* multi-particle state (Fock space state) in quantum field theory transforms according to the rule,

$$U(\Lambda, a)\Psi_{p_1\sigma_1 n_1; p_2\sigma_2 n_2; \cdots}$$
$$= e^{-ia_{\mu}\left[(\Lambda p_1)^{\mu} + (\Lambda p_2)^{\mu} + \cdots\right]} \sqrt{\frac{(\Lambda p_1)^0 (\Lambda p_2)^0 \cdots}{p_1^0 p_2^0 \cdots}} \left(\sum_{\sigma_1' \sigma_2' \cdots} D^{(j_1)}_{\sigma_1' \sigma_1} [W(\Lambda, p_1)] D^{(j_2)}_{\sigma_2' \sigma_2} [W(\Lambda, p_2)] \cdots \right) \Psi_{\Lambda p_1 \sigma_1' n_1; \Lambda p_2 \sigma_2' n_2; \cdots},$$

where $W(\Lambda, p)$ is the Wigner rotation and $D^{(j)}$ is the $(2j + 1)$-dimensional representation of SO(3).

Time Dilation

Time dilation, in the theory of special relativity, the "slowing down" of a clock as determined by an observer who is in relative motion with respect to that clock. In special relativity, an observer in inertial(i.e., nonaccelerating) motion has a well-defined means of determining which events occur simultaneously with a given event. A second inertial observer, who is in relative motion with respect to the first, however, will disagree with the first observer regarding which events are simultaneous with that given event. (Neither observer is wrong in this determination; rather, their disagreement merely reflects the fact that simultaneity is an observer-dependent notion in special relativity.) A notion of simultaneity is required in order to make a comparison of the rates of clocks carried by the two observers. If the first observer's notion of simultaneity is used, it is found that the secondobserver's clock runs slower than the first observer's by a factor of Square root of $\sqrt{(1-v^2/c^2)}$, where υ is the relative velocity of the observers and c equals 299, 792 km (186, 282 miles) per second—i.e., the speed of light. Similarly, using the second observer's notion of simultaneity, it is found that the first observer's clock runs slower by the same factor. Thus, each inertial observer determines that all clocks in motion relative to that observer run slower than that observer's own clock.

A closely related phenomenon predicted by special relativity is the so-called twin paradox. Suppose one of two twins carrying a clock departs on a rocket ship from the other twin, an inertial observer, at a certain time, and they rejoin at a later time. In accordance with the time-dilation effect, the elapsed time on the clock of the twin on the rocket ship will be smaller than that of the inertial observer twin—i.e., the non-inertial twin will have aged less than the inertial observer twin when they rejoin.

The time-dilation effect predicted by special relativity has been accurately confirmed by observations of the increased lifetime of unstable elementary particles traveling at nearly the speed of light. The clock paradox effect also has been substantiated by experiments comparing the elapsed time of an atomic clock on Earth with that of an atomic clock flown in an airplane. The latter experiments, furthermore, have confirmed a gravitational contribution to time dilation, as predicted by the theory of general relativity.

The equation for calculating time dilation is as follows:

$$t = t_0/(1-v^2/c^2)^{1/2}$$

where: t = time observed in the other reference frame.

t_0 = time in observers own frame of reference (rest time).

v = the speed of the moving object.

c = the speed of light in a vacuum.

so in our problem we will let $v = .95c$, $t_0 = 10$ years and we will solve for t which is the time that the earth bound brother measures.

$$t = 10/(1-(.95c)^2/c^2)^{1/2}$$

$$t = 10/(1-.95^2)^{1/2}$$

$$t = 10/.312$$

t = 32 years (the time the earth bound brother measures)

Now let's have a closer look at the equation and determine just what impact the speed of the object has on time dilation. We can see that is the velocity is small compared to the speed of light the quantity v^2/c^2 approaches 0 and the equation simplifies t_0: $t = t_0/1$ which is simply t. So at relatively slow speeds (our everyday speeds) time dilation is not a factor and Newton's Laws are still applicable. Now let's look at high speeds (close to the speed of light), from the equation that as velocity increases the quantity v^2/c^2approaches 1 (but will never quit reach it), causing the quantity$(1-v^2/c^2)^{1/2}$ t_0 become smaller and smaller....therefore causing the time measured by the other observer t_0 become greater thus making our time appear slower (refer back to the example). I know its so confusing!!! read it again, think about it, then study the graph below. As one can see in the graph time dilation starts t_0 "show up" between .4c and .5c. Also notice that the closer one gets to the speed of light the greater impact speed has on time dilation (notice how steep the curve gets towards the end).

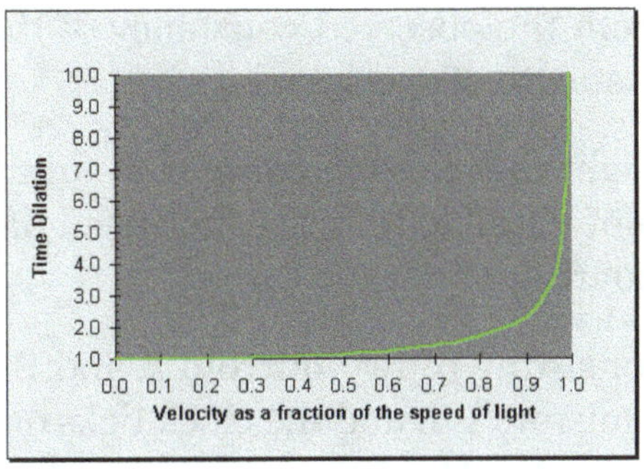

Velocity Time Dilation

Special relativity indicates that, for an observer in an inertial frame of reference, a clock that is moving relative to him will be measured to tick slower than a clock that is at rest in his frame of reference. This case is sometimes called special relativistic time dilation. The faster the relative velocity, the greater the time dilation between one another, with the rate of time reaching zero as one approaches the speed of light (299, 792, 458 m/s). This causes massless particles that travel at the speed of light to be unaffected by the passage of time.

Theoretically, time dilation would make it possible for passengers in a fast-moving vehicle to advance further into the future in a short period of their own time. For sufficiently high speeds, the effect is dramatic. For example, one year of travel might correspond to ten years on Earth. Indeed, a constant 1 g acceleration would permit humans to travel through the entire known Universe in one human lifetime.. At a constant 1 g traveling up to 0.99999999 c it would take 30 years to reach the edge of the universe 13.5 billions lightyears away. Space travelers could then return to Earth billions of years in the future. A scenario based on this idea was presented in the novel *Planet of the Apes* by Pierre Boulle, and the Orion Project has been an attempt toward this idea.

With current technology severely limiting the velocity of space travel, however, the differences experienced in practice are minuscule: after 6 months on the International Space Station (ISS) (which orbits Earth at a speed of about 7, 700 m/s) an astronaut would have aged about 0.005 seconds less than those on Earth. The current human time travel record holder is Russian cosmonaut Sergei Krikalev. He gained 22.68 milliseconds of lifetime on his journeys to space and therefore beat the previous record of about 20 milliseconds by cosmonaut Sergei Avdeyev.

Simple Inference of Velocity Time Dilation

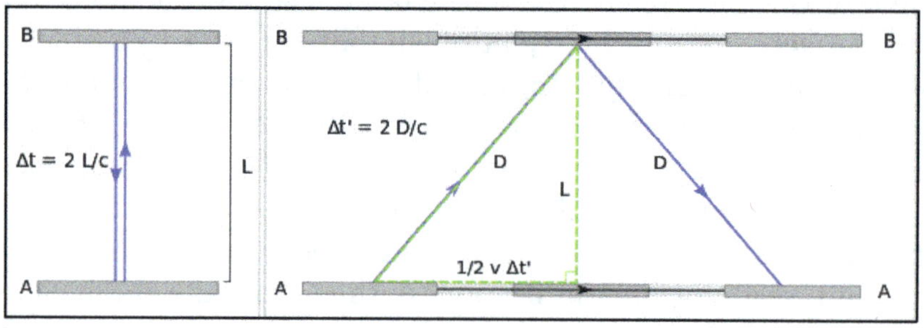

Left: Observer at rest measures time $2L/c$ between co-local events of light signal generation at A and arrival at A. Right: Events according to an observer moving to the left of the setup: bottom mirror A when signal is generated at time $t'=0$, top mirror B when signal gets reflected at time $t'=D/c$, bottom mirror A when signal returns at time $t'=2D/c$.

Time dilation can be inferred from the observed constancy of the speed of light in all reference frames dictated by the second postulate of special relativity.

This constancy of the speed of light means that, counter to intuition, speeds of material objects and light are not additive. It is not possible to make the speed of light appear greater by moving towards or away from the light source.

Consider then, a simple clock consisting of two mirrors A and B, between which a light pulse is bouncing. The separation of the mirrors is L and the clock ticks once each time the light pulse hits either of the mirrors.

In the frame in which the clock is at rest (diagram on the left), the light pulse traces out a path of length $2L$ and the period of the clock is $2L$divided by the speed of light:

$$\Delta t = \frac{2L}{c}.$$

From the frame of reference of a moving observer traveling at the speed v relative to the resting frame of the clock (diagram at right), the light pulse is seen as tracing out a longer, angled path. Keeping the speed of light constant for all inertial observers, requires a lengthening of the period of this clock from the moving observer's perspective. That is to say, in a frame moving relative to the local clock, this clock will appear to be running more slowly. Straightforward application of the Pythagorean theorem leads to the well-known prediction of special relativity:

The total time for the light pulse to trace its path is given by,

$$\Delta t' = \frac{2D}{c}.$$

The length of the half path can be calculated as a function of known quantities as,

$$D = \sqrt{\left(\frac{1}{2} v \Delta t'\right)^2 + L^2}.$$

Elimination of the variables D and L from these three equations results in,

$$\Delta t' = \frac{\Delta t}{\sqrt{1-\frac{v^2}{c^2}}},$$

which expresses the fact that the moving observer's period of the clock Ät′ is longer than the period Δt in the frame of the clock itself.

Reciprocity

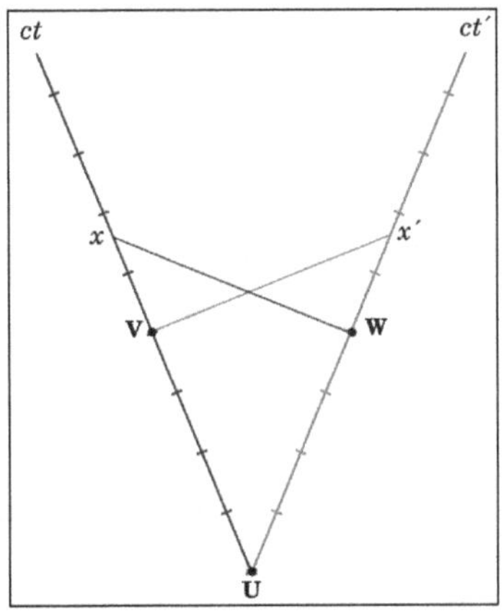

Time UV of a clock in S is shorter compared to Ux′ in S′, and time UW of a clock in S′ is shorter compared to Ux in S.

Given a certain frame of reference, and the "stationary" observer, if a second observer accompanied the "moving" clock, each of the observers would perceive the other's clock as ticking at a *slower* rate than their own local clock, due to them both perceiving the other to be the one that's in motion relative to their own stationary frame of reference.

Common sense would dictate that, if the passage of time has slowed for a moving object, said object would observe the external world's time to be correspondingly sped up. Counterintuitively, special relativity predicts the opposite. When two observers are in motion relative to each other, each will measure the other's clock slowing down, in concordance with them being moving relative to the observer's frame of reference.

While this seems self-contradictory, a similar oddity occurs in everyday life. If two persons A and B observe each other from a distance, B will appear small to A, but at the same time A will appear small to B. Being familiar with the effects of perspective, there is no contradiction or paradox in this situation.

The reciprocity of the phenomenon also leads to the so-called twin paradox where the aging of twins, one staying on Earth and the other embarking on a space travel, is compared, and where the reciprocity suggests that both persons should have the same age when they reunite. On the contrary, at the end of the round-trip, the traveling twin will be younger than his brother on Earth. The dilemma posed by the paradox, however, can be explained by the fact that the traveling twin must markedly accelerate in at least three phases of the trip (beginning, direction change, and end), while the other will only experience negligible acceleration, due to rotation and revolution of Earth. During the acceleration phases of the space travel, time dilation is not symmetric.

Experimental Testing

Doppler Effect

- Ives and Stilwell (1938, 1941) stated purpose of these experiments was to verify the time dilation effect, predicted by Larmor–Lorentz ether theory, due to motion through the ether using Einstein's suggestion that Doppler effect in canal rays would provide a suitable experiment. These experiments measured the Doppler shift of the radiation emitted from cathode rays, when viewed from directly in front and from directly behind. The high and low frequencies detected were not the classically predicted values,

  $$\frac{f_o}{1-v/c} \quad \text{and} \quad \frac{f_o}{1+v/c}.$$

  The high and low frequencies of the radiation from the moving sources were measured as,

  $$\sqrt{\frac{1+v/c}{1-v/c}}f_o = \gamma(1+v/c)f_o \quad \text{and} \quad \sqrt{\frac{1-v/c}{1+v/c}}f_o = \gamma(1-v/c)f_o,$$

  as deduced by Einstein (1905) from the Lorentz transformation, when the source is running slow by the Lorentz factor.

- Hasselkamp, Mondry, and Scharmann (1979) measured the Doppler shift from a source moving at right angles to the line of sight. The most general relationship between frequencies of the radiation from the moving sources is given by:

  $$f_{\text{detected}} = f_{\text{rest}}\left(1-\frac{v}{c}\cos\phi\right)/\sqrt{1-v^2/c^2}$$

  as deduced by Einstein (1905). For ϕ = 90° (cos ϕ = 0) this reduces to fdetected = frestγ. This lower frequency from the moving source can be attributed to the time dilation effect and is often called the transverse Doppler effect and was predicted by relativity.

- In 2010 time dilation was observed at speeds of less than 10 meters per second using optical atomic clocks connected by 75 meters of optical fiber.

Moving Particles

- A comparison of muon lifetimes at different speeds is possible. In the laboratory, slow muons are produced; and in the atmosphere, very fast moving muons are introduced by cosmic rays. Taking the muon lifetime at rest as the laboratory value of 2.197 μs, the lifetime of a cosmic ray produced muon traveling at 98% of the speed of light is about five times longer, in agreement with observations. An example is Rossi and Hall (1941), who compared the population of cosmic-ray-produced muons at the top of a mountain to that observed at sea level.

- The lifetime of particles produced in particle accelerators appears longer due to time dilation. In such experiments the "clock" is the time taken by processes leading to muon decay, and these processes take place in the moving muon at its own "clock rate", which is much slower than the laboratory clock. This is routinely taken into account in particle physics, and many dedicated measurements have been performed. For instance, in the muon storage ring at CERN the lifetime of muons circulating with $\gamma = 29.327$ was found to be dilated to 64.378 μs, confirming time dilation to an accuracy of 0.9 ± 0.4 parts per thousand.

Proper Time and Minkowski Diagram

Minkowski Diagram and Twin Paradox

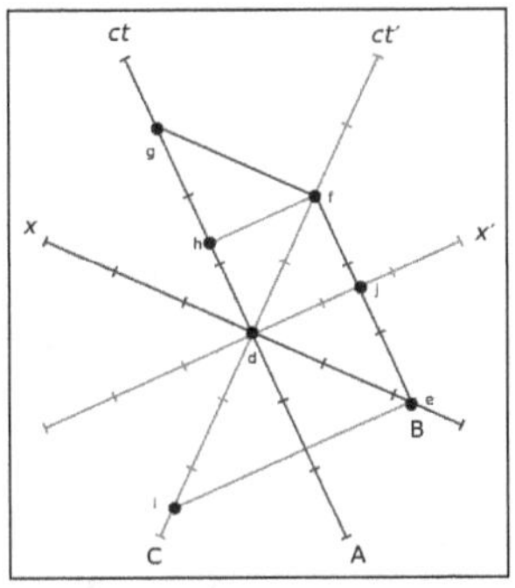

Clock C in relative motion between two synchronized clocks A and B. C meets A at *d*, and B at *f*.

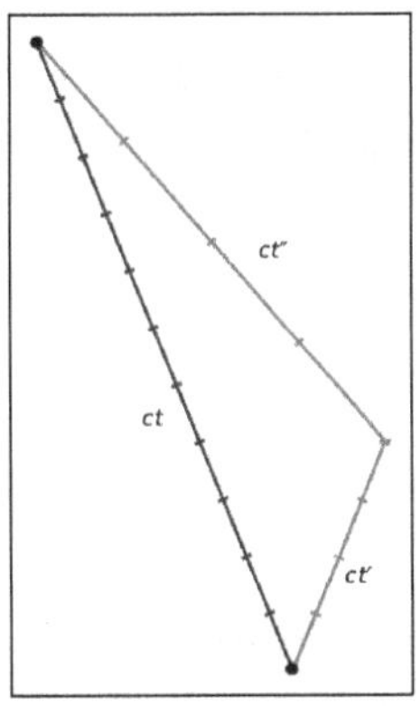

Twin paradox. One twin has to change frames, leading to different proper times in the twin's world lines.

In the Minkowski diagram from the second image on the right, clock C resting in inertial frame S′ meets clock A at *d* and clock B at *f* (both resting in S). All three clocks simultaneously start to tick in S. The worldline of A is the ct-axis, the worldline of B intersecting *f* is parallel to the ct-axis, and the worldline of C is the ct′-axis. All events simultaneous with *d* in S are on the x-axis, in S′ on the x′-axis.

The proper time between two events is indicated by a clock present at both events. It is invariant, i.e., in all inertial frames it is agreed that this time is indicated by that clock. Interval *df* is therefore the proper time of clock C, and is shorter with respect to the coordinate times *ef=dg* of clocks B and A in S. Conversely, also proper time *ef* of B is shorter with respect to time *if* in S′, because event *e* was measured in S′ already at time *i*due to relativity of simultaneity, long before C started to tick.

From that it can be seen, that the proper time between two events indicated by an unaccelerated clock present at both events, compared with the synchronized coordinate time measured in all other inertial frames, is always the *minimal* time interval between those events. However, the interval between two events can also correspond to the proper time of accelerated clocks present at both events. Under all possible proper times between two events, the proper time of the unaccelerated clock is *maximal*, which is the solution to the twin paradox.

Derivation and Formulation

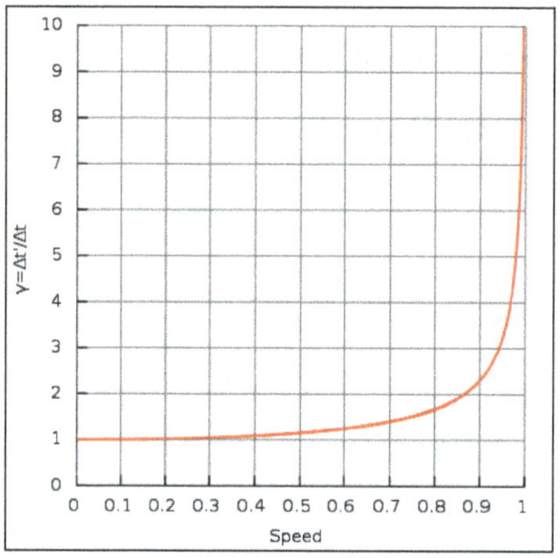

Lorentz factor as a function of speed (in natural units where c = 1). Notice that for small speeds (less than 0.1), γ is approximately 1.

In addition to the light clock used above, the formula for time dilation can be more generally derived from the temporal part of the Lorentz transformation. Let there be two events at which the moving clock indicates t_a and t_b, thus

$$t_a' = \frac{t_a - \frac{vx_a}{c^2}}{\sqrt{1-\frac{v^2}{c^2}}}, t_b' = \frac{t_b - \frac{vx_b}{c^2}}{\sqrt{1-\frac{v^2}{c^2}}}.$$

Since the clock remains at rest in its inertial frame, it follows $x_a = x_b$, thus the interval $\Delta t' = t_b' - t_a'$ is given by,

$$\Delta t' = \gamma \Delta t = \frac{\Delta t}{\sqrt{1-\frac{v^2}{c^2}}}$$

where Δt is the time interval between *two co-local events* (i.e. happening at the same place) for an observer in some inertial frame (e.g. ticks on his clock), known as the *proper time*, $\Delta t'$ is the time interval between those same events, as measured by another observer, inertially moving with velocity v with respect to the former observer, v is the relative velocity between the observer and the moving clock, c is the speed of light, and the Lorentz factor (conventionally denoted by the Greek letter gamma or γ) is,

$$\gamma = \frac{1}{\sqrt{1-\frac{v^2}{c^2}}}.$$

Thus the duration of the clock cycle of a moving clock is found to be increased: it is measured to be "running slow". The range of such variances in ordinary life, where $v \ll c$, even considering space travel, are not great enough to produce easily detectable time dilation effects and such vanishingly small effects can be safely ignored for most purposes. It is only when an object approaches speeds on the order of 30, 000 km/s (1/10 the speed of light) that time dilation becomes important.

Hyperbolic Motion

In special relativity, time dilation is most simply described in circumstances where relative velocity is unchanging. Nevertheless, the Lorentz equations allow one to calculate proper time and movement in space for the simple case of a spaceship which is applied with a force per unit mass, relative to some reference object in uniform (i.e. constant velocity) motion, equal to g throughout the period of measurement.

Let t be the time in an inertial frame subsequently called the rest frame. Let x be a spatial coordinate, and let the direction of the constant acceleration as well as the spaceship's velocity (relative to the rest frame) be parallel to the x-axis. Assuming the spaceship's position at time $t = 0$ being $x = 0$ and the velocity being v_0 and defining the following abbreviation,

$$\gamma_0 = \frac{1}{\sqrt{1-v_0^2/c^2}},$$

the following formulas hold:

Position:

$$x(t) = \frac{c^2}{g}\left(\sqrt{1+\frac{(gt+v_0\gamma_0)^2}{c^2}} - \gamma_0\right).$$

Velocity:

$$v(t) = \frac{gt+v_0\gamma_0}{\sqrt{1+\frac{(gt+v_0\gamma_0)^2}{c^2}}}.$$

Proper time as function of coordinate time:

$$\tau(t) = \tau_0 + \int_0^t \sqrt{1-\left(\frac{v(t')}{c}\right)^2}\, dt'.$$

In the case where $v(0) = v_0 = 0$ and $\tau(0) = \tau_0 = 0$ the integral can be expressed as a logarithmic function or, equivalently, as an inverse hyperbolic function:

$$\tau(t) = \frac{c}{g} ln\left(\frac{gt}{c} + \sqrt{1+\left(\frac{gt}{c}\right)^2}\right) = \frac{c}{g} arsinh\left(\frac{gt}{c}\right).$$

As functions of the proper time τ of the ship, the following formulae hold:

Position:

$$x(\tau) = \frac{c^2}{g}\left(\cosh\frac{g\tau}{c} - 1\right).$$

Velocity:

$$v(\tau) = c\tanh\frac{g\tau}{c}.$$

Coordinate time as function of proper time:

$$t(\tau) = \frac{c}{g}\sinh\frac{g\tau}{c}.$$

Clock Hypothesis

The clock hypothesis is the assumption that the rate at which a clock is affected by time dilation does not depend on its acceleration but only on its instantaneous velocity. This is equivalent to stating that a clock moving along a path P measures the proper time, defined by:

$$d\tau = \int_P \sqrt{dt^2 - dx^2/c^2 - dy^2/c^2 - dz^2/c^2}.$$

The clock hypothesis was implicitly (but not explicitly) included in Einstein's original 1905 formulation of special relativity. Since then, it has become a standard assumption and is usually included in the axioms of special relativity, especially in the light of experimental verification up to very high accelerations in particle accelerators.

Gravitational time dilation is a phenomenon whereby time runs slower when in a higher gravitational potential. Put simply, the closer you are to a large body like the Earth the slower time runs, thus time runs slower for someone on the surface of the earth compared to someone in orbit around the earth.

Whilst the Einstein's field equation for General relativity is very complicated the equation for time dilation is much simpler and bears certain resemblance to the equation for time dilation for relative motion. The equation is:

$$t_o = t_f\sqrt{1 - \frac{2GM}{rc^2}}$$

In this equation to is the proper time for an event, i.e. the time measured when observer and event are in the same gravitational potential and tf is the time as measured when at an infinite distance from any mass. The values G and c are again the Newtonian gravitational potential and speed of light respectively and M and r are the mass of the object you are near and r is your distance from said object respectively.

This equation can be written differently by applying Schwarzschild's solution to general relativity which gives you the Schwarzschild radius; the radius of an object of mass M, for which if all mass is contained within that radius then the gravitational field is too strong that not even light can escape.

$$r_s = \frac{2GM}{c^2} \Rightarrow t_o = t_f\sqrt{1-\frac{2GM}{rc^2}} = t_f\sqrt{1-\frac{r_s}{r}}$$

If one therefore applies this formula whose mass is contained within their Schwarzschild radius, i.e. Black Holes, then when something or someone reaches the edge of a black hole when r=rs time for the object/person appears to stop from the point of view of an observer further away. Thus you can never see something fall into a black hole as time stops at the edge of the black hole.

It then gets even stranger once the person has fallen into the black hole (although you will never see it) as the value in the square root operator becomes negative so the answer come back as an imaginary number and no one knows how to interpret an imaginary result to the real world.

References

- Bartusiak, Marcia (2000), Einstein's Unfinished Symphony: Listening to the Sounds of Space-Time, Berkley, ISBN 978-0-425-18620-6
- Lorentz-transformations, science: britannica.com, Retrieved 08,July 2020
- Font, José A. (2003), "Numerical Hydrodynamics in General Relativity", Living Reviews in Relativity, 6, Bibcode:2003LRR.....6....4F, doi:10.12942/lrr-2003-4, PMC 5256036, PMID 28179854, retrieved 19,August 2020
- What is relativistic mass and why it is not used much: physicsforums com Retrieved 18,May 2020
- Blanchet, Luc (2006), "Gravitational Radiation from Post-Newtonian Sources and Inspiralling Compact Binaries", Living Reviews in Relativity, 9, Bibcode:2006LRR.....9....4B, doi:10.12942/lrr-2006-4, retrieved 09, August 2020
- General-relativity: thestargarden.co.uk, Retrieved 29,June 2020
- Physics, relativistic-mass-formula-546: softschools.com, Retrieved 11,April 2020
- Theory of relativity: allaboutscience.org, Retrieved 09, July 2020
- What-is-gravitational-time-dilation: users.sussex.ac.uk, Retrieved 18,April 2020
- Physics, relativistic-energy-formula-545: softschools.com, Retrieved 28,June 2020
- Relativistic-mechanics, science: britannica.com, Retrieved 29,June 2020
- Time-dilation, science: britannica.com, Retrieved 27,July 2020
- Derivation-of-relativistic-mass: thecuriousastronomer.wordpress.com, Retrieved 22,July 2020

6

Important Principles and Laws used in Physics

Physics can be understood through various theories, principles and laws such as classical Hamiltonian mechanics, Archimedes' principle, Bernoulli's principle, physical law, conservation law, inverse-square law, Stokes' law, Faxen's law and Hagen-Poiseuille equation. This chapter has been carefully written to provide an easy understanding of these various theories, principles and laws of physics.

Archimedes' Principle

In *On Floating Bodies*, Archimedes suggested that (c. 250 BC):

Any object, wholly or partially immersed in a fluid, is buoyed up by a force equal to the weight of the fluid displaced by the object.

Practically, the Archimedes' principle allows the buoyancy of an object partially or fully immersed in a liquid to be calculated. The downward force on the object is simply its weight. The upward, or buoyant force on the object is that stated by Archimedes' principle, above. Thus the net upward force on the object is the difference between the buoyant force and its weight. If this net force is positive, the object rises; if negative, the object sinks; and if zero, the object is neutrally buoyant - that is, it remains in place without either rising or sinking. In simple words, Archimedes' principle states that when a body is partially or completely immersed in a fluid, it experiences an apparent loss in weight which is equal to the weight of the fluid displaced by the immersed part of the body.

Formula

Consider a cube immersed in a fluid, with its sides parallel to the direction of gravity. The fluid will exert a normal force on each face, and therefore only the forces on the top and bottom faces will contribute to buoyancy. The pressure difference between the bottom and the top face is directly proportional to the height (difference in depth). Multiplying the pressure difference by the area of a face gives the net force on the cube – the buoyancy, or the weight of the fluid displaced. By extending this reasoning to irregular shapes, we can see that, whatever the shape of the submerged body, the buoyant force is equal to the weight of the fluid displaced. apparent loss in wt of water= wt of object in air-wt of object in water.

The weight of the displaced fluid is directly proportional to the volume of the displaced fluid (if the surrounding fluid is of uniform density). The weight of the object in the fluid is reduced, because of the force acting on it, which is called upthrust. In simple terms, the principle states that the buoyant force on an object is equal to the weight of the fluid displaced by the object, or the density of the fluid multiplied by the submerged volume times the gravitational constant, g. Thus, among completely submerged objects with equal masses, objects with greater volume have greater buoyancy.

Suppose a rock's weight is measured as 10 newtons when suspended by a string in a vacuum with gravity acting on it. Suppose that when the rock is lowered into water, it displaces water of weight 3 newtons. The force it then exerts on the string from which it hangs would be 10 newtons minus the 3 newtons of buoyant force: 10 − 3 = 7 newtons. Buoyancy reduces the apparent weight of objects that have sunk completely to the sea floor. It is generally easier to lift an object up through the water than it is to pull it out of the water.

For a fully submerged object, Archimedes' principle can be reformulated as follows:

$$\text{apparent immersed weight} = \text{weight of object} - \text{weight of displaced fluid}$$

then inserted into the quotient of weights, which has been expanded by the mutual volume,

$$\frac{\text{density of object}}{\text{density of fluid}} = \frac{\text{weight}}{\text{weight of displaced fluid}}$$

yields the formula below. The density of the immersed object relative to the density of the fluid can easily be calculated without measuring any volumes is,

$$\frac{\text{density of object}}{\text{density of fluid}} = \frac{\text{weight}}{\text{weight} - \text{apparent immersed weight}}.$$

(This formula is used for example in describing the measuring principle of a dasymeter and of hydrostatic weighing.)

Example: If you drop wood into water, buoyancy will keep it afloat.

Example: A helium balloon in a moving car. When increasing speed or driving in a curve, the air moves in the opposite direction to the car's acceleration. However, due to buoyancy, the balloon is pushed "out of the way" by the air, and will actually drift in the same direction as the car's acceleration.

When an object is immersed in a liquid, the liquid exerts an upward force, which is known as the buoyant force, that is proportional to the weight of the displaced liquid. The sum force acting on the object, then, is equal to the difference between the weight of the object ('down' force) and the weight of displaced liquid ('up' force). Equilibrium, or neutral buoyancy, is achieved when these two weights (and thus forces) are equal.

Refinements

Archimedes' principle does not consider the surface tension (capillarity) acting on the body. Moreover, Archimedes' principle has been found to break down in complex fluids.

Principle of Flotation

Archimedes' principle shows buoyant force and displacement of fluid. However, the concept of Archimedes' principle can be applied when considering why objects float. Proposition 5 of Archimedes' treatise *On Floating Bodies* states that:

Any floating object displaces its own weight of fluid.

—*Archimedes of Syracuse*

In other words, for an object floating on a liquid surface (like a boat) or floating submerged in a fluid (like a submarine in water or dirigible in air) the weight of the displaced liquid equals the weight of the object. Thus, only in the special case of floating does the buoyant force acting on an object equal the objects weight. Consider a 1-ton block of solid iron. As iron is nearly eight times denser than water, it displaces only 1/8 ton of water when submerged, which is not enough to keep it afloat. Suppose the same iron block is reshaped into a bowl. It still weighs 1 ton, but when it is put in water, it displaces a greater volume of water than when it was a block. The deeper the iron bowl is immersed, the more water it displaces, and the greater the buoyant force acting on it. When the buoyant force equals 1 ton, it will sink no farther.

When any boat displaces a weight of water equal to its own weight, it floats. This is often called the "principle of flotation": A floating object displaces a weight of fluid equal to its own weight. Every ship, submarine, and dirigible must be designed to displace a weight of fluid at least equal to its own weight. A 10,000-ton ship must be built wide enough to displace 10,000 tons of water before it sinks too deep in the water. The same is true for vessels in air: a dirigible that weighs 100 tons needs to displace 100 tons of air. If it displaces more, it rises; if it displaces less, it falls. If the dirigible displaces exactly its weight, it hovers at a constant altitude.

It is important to realize that, while they are related to it, the principle of flotation and the concept that a submerged object displaces a volume of fluid equal to its own volume are *not* Archimedes' principle. Archimedes' principle, as stated above, equates the *buoyant force* to the weight of the fluid displaced.

One common point of confusion regarding Archimedes' principle is the meaning of displaced volume. Common demonstrations involve measuring the rise in water level when an object floats on the surface in order to calculate the displaced water.This measurement approach fails with a buoyant submerged object because the rise in the water level is directly related to the volume of the object and not the mass (except if the effective density of the object equals exactly the fluid density). Instead, in the case of submerged buoyant objects, the whole volume of fluid directly above the sample should be considered as the displaced volume. Another common point of confusion regarding Archimedes' principle is that it only applies to submerged objects that are buoyant, not sunk objects. In the case of a sunk object the mass of displaced fluid is less than the mass of the object and the difference is associated with the object's potential energy.

Bernoulli's Principle

In fluid dynamics, Bernoulli's principle states that an increase in the speed of a fluid occurs simultaneously with a decrease in pressure or a decrease in the fluid's potential energy. The principle is named after Daniel Bernoulli who published it in his book *Hydrodynamica* in 1738.

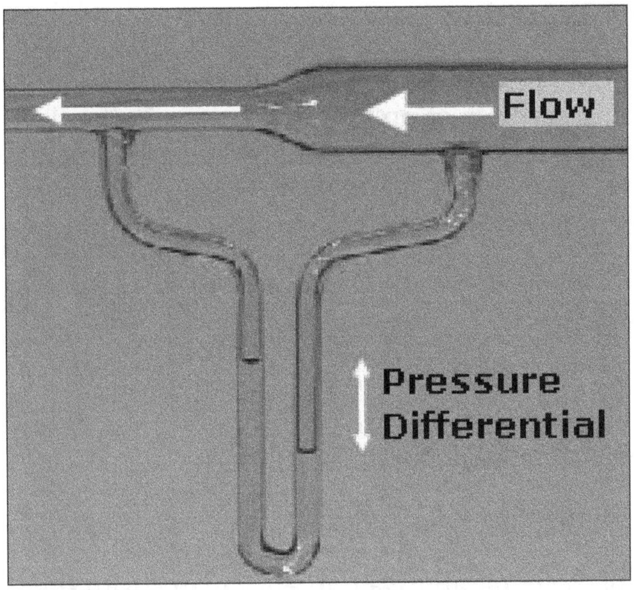

A flow of water into a venturi meter. The kinetic energy increases at the expense of the fluid pressure, as shown by the difference in height of the two columns of water.

Bernoulli's principle can be applied to various types of fluid flow, resulting in various forms of Bernoulli's equation; there are different forms of Bernoulli's equation for different types of flow. The simple form of Bernoulli's equation is valid for incompressible flows (e.g. most liquid flows and gases moving at low Mach number). More advanced forms may be applied to compressible flows at higher Mach numbers.

Bernoulli's principle can be derived from the principle of conservation of energy. This states that, in a steady flow, the sum of all forms of energy in a fluid along a streamline is the same at all points on that streamline. This requires that the sum of kinetic energy, potential energy and internal energy remains constant. Thus an increase in the speed of the fluid – implying an increase in both its dynamic pressure and kinetic energy – occurs with a simultaneous decrease in (the sum of) its static pressure, potential energy and internal energy. If the fluid is flowing out of a reservoir, the sum of all forms of energy is the same on all streamlines because in a reservoir the energy per unit volume (the sum of pressure and gravitational potential $\rho\, g\, h$) is the same everywhere.

Bernoulli's principle can also be derived directly from Newton's 2nd law. If a small volume of fluid is flowing horizontally from a region of high pressure to a region of low pressure, then there is more pressure behind than in front. This gives a net force on the volume, accelerating it along the streamline.

Fluid particles are subject only to pressure and their own weight. If a fluid is flowing horizontally and along a section of a streamline, where the speed increases it can only be because the fluid on that section has moved from a region of higher pressure to a region of lower pressure; and if its speed decreases, it can only be because it has moved from a region of lower pressure to a region of higher pressure. Consequently, within a fluid flowing horizontally, the highest speed occurs where the pressure is lowest, and the lowest speed occurs where the pressure is highest.

Incompressible Flow Equation

In most flows of liquids, and of gases at low Mach number, the density of a fluid parcel can be considered to be constant, regardless of pressure variations in the flow. Therefore, the fluid can be considered to be incompressible and these flows are called incompressible flows. Bernoulli performed his experiments on liquids, so his equation in its original form is valid only for incompressible flow. A common form of Bernoulli's equation, valid at any arbitrary point along a streamline, is:

$$\frac{v^2}{2} + gz + \frac{p}{\rho} = \text{constant} \qquad (A)$$

where:

v is the fluid flow speed at a point on a streamline,

g is the acceleration due to gravity,

z is the elevation of the point above a reference plane, with the positive z-direction pointing upward – so in the direction opposite to the gravitational acceleration,

p is the pressure at the chosen point, and

ρ is the density of the fluid at all points in the fluid.

The constant on the right-hand side of the equation depends only on the streamline chosen, whereas v, z and p depend on the particular point on that streamline.

The following assumptions must be met for this Bernoulli equation to apply:

- The flow must be steady, i.e. the fluid velocity at a point cannot change with time,
- The flow must be incompressible – even though pressure varies, the density must remain constant along a streamline;
- Friction by viscous forces has to be negligible.

For conservative force fields (not limited to the gravitational field), Bernoulli's equation can be generalized as:

$$\frac{v^2}{2} + \Psi + \frac{p}{\rho} = \text{constant}$$

where Ψ is the force potential at the point considered on the streamline. E.g. for the Earth's gravity $\Psi = gz$.

By multiplying with the fluid density ρ, equation (A) can be rewritten as:

$$\tfrac{1}{2}\rho v^2 + \rho g z + p = \text{constant}$$

or:

$$q + \rho g h = p_0 + \rho g z = \text{constant}$$

where:

$q=\frac{1}{2}\rho v^2$ is dynamic pressure,

$h=z+\frac{p}{\rho g}$ is the piezometric head or hydraulic head (the sum of the elevation z and the pressure head) and

$p_0=p+q$ is the total pressure (the sum of the static pressure p and dynamic pressure q).

The constant in the Bernoulli equation can be normalised. A common approach is in terms of total head or energy head H:

$$H=z+\frac{p}{\rho g}+\frac{v^2}{2g}=h+\frac{v^2}{2g},$$

The above equations suggest there is a flow speed at which pressure is zero, and at even higher speeds the pressure is negative. Most often, gases and liquids are not capable of negative absolute pressure, or even zero pressure, so clearly Bernoulli's equation ceases to be valid before zero pressure is reached. In liquids – when the pressure becomes too low – cavitation occurs. The above equations use a linear relationship between flow speed squared and pressure. At higher flow speeds in gases, or for sound waves in liquid, the changes in mass density become significant so that the assumption of constant density is invalid.

Simplified Form

In many applications of Bernoulli's equation, the change in the $\rho g z$ term along the streamline is so small compared with the other terms that it can be ignored. For example, in the case of aircraft in flight, the change in height z along a streamline is so small the $\rho g z$ term can be omitted. This allows the above equation to be presented in the following simplified form:

$$p+q=p_0$$

where p_0 is called 'total pressure', and q is 'dynamic pressure'. Many authors refer to the pressure p as static pressure to distinguish it from total pressure p_0 and dynamic pressure q. In *Aerodynamics*, L.J. Clancy writes: "To distinguish it from the total and dynamic pressures, the actual pressure of the fluid, which is associated not with its motion but with its state, is often referred to as the static pressure, but where the term pressure alone is used it refers to this static pressure."

The simplified form of Bernoulli's equation can be summarized in the following memorable word equation:

static pressure + dynamic pressure = total pressure

Every point in a steadily flowing fluid, regardless of the fluid speed at that point, has its own unique static pressure p and dynamic pressure q. Their sum $p + q$ is defined to be the total pressure p_0. The significance of Bernoulli's principle can now be summarized as *total pressure is constant along a streamline*.

If the fluid flow is irrotational, the total pressure on every streamline is the same and Bernoulli's principle can be summarized as *total pressure is constant everywhere in the fluid flow*. It is reasonable to assume that irrotational flow exists in any situation where a large body of fluid is flowing past a solid body. Examples are aircraft in flight, and ships moving in open bodies of water. However, it is important to remember that Bernoulli's principle does not apply in the boundary layer or in fluid flow through long pipes.

If the fluid flow at some point along a stream line is brought to rest, this point is called a stagnation point, and at this point the total pressure is equal to the stagnation pressure.

Applicability of Incompressible Flow Equation to Flow of Gases

Bernoulli's equation is sometimes valid for the flow of gases: provided that there is no transfer of kinetic or potential energy from the gas flow to the compression or expansion of the gas. If both the gas pressure and volume change simultaneously, then work will be done on or by the gas. In this case, Bernoulli's equation – in its incompressible flow form – cannot be assumed to be valid. However, if the gas process is entirely isobaric, or isochoric, then no work is done on or by the gas, (so the simple energy balance is not upset). According to the gas law, an isobaric or isochoric process is ordinarily the only way to ensure constant density in a gas. Also the gas density will be proportional to the ratio of pressure and absolute temperature, however this ratio will vary upon compression or expansion, no matter what non-zero quantity of heat is added or removed. The only exception is if the net heat transfer is zero, as in a complete thermodynamic cycle, or in an individual isentropic (frictionless adiabatic) process, and even then this reversible process must be reversed, to restore the gas to the original pressure and specific volume, and thus density. Only then is the original, unmodified Bernoulli equation applicable. In this case the equation can be used if the flow speed of the gas is sufficiently below the speed of sound, such that the variation in density of the gas (due to this effect) along each streamline can be ignored. Adiabatic flow at less than Mach 0.3 is generally considered to be slow enough.

Unsteady Potential Flow

The Bernoulli equation for unsteady potential flow is used in the theory of ocean surface waves and acoustics.

For an irrotational flow, the flow velocity can be described as the gradient $\nabla\varphi$ of a velocity potential φ. In that case, and for a constant density ρ, the momentum equations of the Euler equations can be integrated to:

$$\frac{\partial\varphi}{\partial t}+\frac{1}{2}v^2+\frac{p}{\rho}+gz=f(t),$$

which is a Bernoulli equation valid also for unsteady—or time dependent—flows. Here $\partial\varphi/\partial t$ denotes the partial derivative of the velocity potential φ with respect to time t, and $v = |\nabla\varphi|$ is the flow speed. The function $f(t)$ depends only on time and not on position in the fluid. As a result, the Bernoulli equation at some moment t does not only apply along a certain streamline, but in the whole fluid domain. This is also true for the special case of a steady irrotational flow, in which case f is a constant.

Further $f(t)$ can be made equal to zero by incorporating it into the velocity potential using the transformation

$$\Phi = \varphi - \int_{t_0}^{t} f(\tau)\mathrm{d}\tau,$$

resulting in $$\frac{\partial \Phi}{\partial t} + \tfrac{1}{2}v^2 + \frac{p}{\rho} + gz = 0$$

Note that the relation of the potential to the flow velocity is unaffected by this transformation: $\nabla\Phi = \nabla\varphi$.

The Bernoulli equation for unsteady potential flow also appears to play a central role in Luke's variational principle, a variational description of free-surface flows using the Lagrangian.

Compressible Flow Equation

Bernoulli developed his principle from his observations on liquids, and his equation is applicable only to incompressible fluids, and compressible fluids up to approximately Mach number 0.3. It is possible to use the fundamental principles of physics to develop similar equations applicable to compressible fluids. There are numerous equations, each tailored for a particular application, but all are analogous to Bernoulli's equation and all rely on nothing more than the fundamental principles of physics such as Newton's laws of motion or the first law of thermodynamics.

Compressible Flow in Fluid Dynamics

For a compressible fluid, with a barotropic equation of state, and under the action of conservative forces,

$$\frac{v^2}{2} + \int_{p_1}^{p} \frac{d\tilde{p}}{\rho(\tilde{p})} + \Psi = \text{constant} \quad \text{(along a streamline)}$$

where:

p is the pressure.

ρ is the density.

v is the flow speed.

Ψ is the potential associated with the conservative force field, often the gravitational potential.

In engineering situations, elevations are generally small compared to the size of the Earth, and the time scales of fluid flow are small enough to consider the equation of state as adiabatic. In this case, the above equation becomes,

$$\frac{v^2}{2} + gz + \left(\frac{\gamma}{\gamma - 1}\right)\frac{p}{\rho} = \text{constant} \quad \text{(along a streamline)}$$

where, in addition to the terms listed above:

γ is the ratio of the specific heats of the fluid.

g is the acceleration due to gravity.

z is the elevation of the point above a reference plane.

In many applications of compressible flow, changes in elevation are negligible compared to the other terms, so the term gz can be omitted. A very useful form of the equation is then:

$$\frac{v^2}{2}+\left(\frac{\gamma}{\gamma-1}\right)\frac{p}{\rho}=\left(\frac{\gamma}{\gamma-1}\right)\frac{p_0}{\rho_0}$$

where:

p_0 is the total pressure.

ρ_0 is the total density.

Compressible Flow in Thermodynamics

The most general form of the equation, suitable for use in thermodynamics in case of (quasi) steady flow, is:

$$\frac{v^2}{2}+\Psi+w=\text{constant}.$$

Here w is the enthalpy per unit mass, which is also often written as h.

Note that $w=\epsilon+\frac{p}{\rho}$ where ε is the thermodynamic energy per unit mass, also known as the specific internal energy. So, for constant internal energy ε the equation reduces to the incompressible-flow form.

The constant on the right hand side is often called the Bernoulli constant and denoted b. For steady inviscid adiabatic flow with no additional sources or sinks of energy, b is constant along any given streamline. More generally, when b may vary along streamlines, it still proves a useful parameter, related to the "head" of the fluid.

When the change in Ψ can be ignored, a very useful form of this equation is:

$$\frac{v^2}{2}+w=w_0$$

where w_0 is total enthalpy. For a calorically perfect gas such as an ideal gas, the enthalpy is directly proportional to the temperature, and this leads to the concept of the total (or stagnation) temperature.

When shock waves are present, in a reference frame in which the shock is stationary and the flow is steady, many of the parameters in the Bernoulli equation suffer abrupt changes in passing through the shock. The Bernoulli parameter itself, however, remains unaffected. An exception to this rule is radiative shocks, which violate the assumptions leading to the Bernoulli equation, namely the lack of additional sinks or sources of energy.

Derivations of Bernoulli Equation

Bernoulli Equation for Incompressible Fluids

The Bernoulli equation for incompressible fluids can be derived by either integrating Newton's second law of motion or by applying the law of conservation of energy between two sections along a streamline, ignoring viscosity, compressibility, and thermal effects.

Derivation through Integrating Newton's Second Law of Motion

The simplest derivation is to first ignore gravity and consider constrictions and expansions in pipes that are otherwise straight, as seen in Venturi effect. Let the x axis be directed down the axis of the pipe. Define a parcel of fluid moving through a pipe with cross-sectional area A, the length of the parcel is dx, and the volume of the parcel A dx. If mass density is ρ, the mass of the parcel is density multiplied by its volume $m = \rho A$ dx. The change in pressure over distance dx is dp and flow velocity v = dx / dt. Apply Newton's second law of motion (force = mass × acceleration) and recognizing that the effective force on the parcel of fluid is –A dp. If the pressure decreases along the length of the pipe, dp is negative but the force resulting in flow is positive along the x axis.

$$m\frac{\mathrm{d}v}{\mathrm{d}t}=F$$

$$\rho A\,\mathrm{d}x\frac{\mathrm{d}v}{\mathrm{d}t}=-A\,\mathrm{d}p$$

$$\rho\frac{\mathrm{d}v}{\mathrm{d}t}=-\frac{\mathrm{d}p}{\mathrm{d}x}$$

In steady flow the velocity field is constant with respect to time, $v = v(x) = v(x(t))$, so v itself is not directly a function of time t. It is only when the parcel moves through x that the cross sectional area changes: v depends on t only through the cross-sectional position $x(t)$.

$$\frac{\mathrm{d}v}{\mathrm{d}t}=\frac{\mathrm{d}v}{\mathrm{d}x}\frac{\mathrm{d}x}{\mathrm{d}t}=\frac{\mathrm{d}v}{\mathrm{d}x}v=\frac{d}{\mathrm{d}x}\left(\frac{v^2}{2}\right).$$

With density ρ constant, the equation of motion can be written as,

$$\frac{\mathrm{d}}{\mathrm{d}x}\left(\rho\frac{v^2}{2}+p\right)=0$$

by integrating with respect to x,

$$\frac{v^2}{2}+\frac{p}{\rho}=C$$

where C is a constant, sometimes referred to as the Bernoulli constant. It is not a universal constant, but rather a constant of a particular fluid system. The deduction is: where the speed is large, pressure is low and vice versa. In the above derivation, no external work-energy principle is invoked. Rather, Bernoulli's principle was derived by a simple manipulation of Newton's second law.

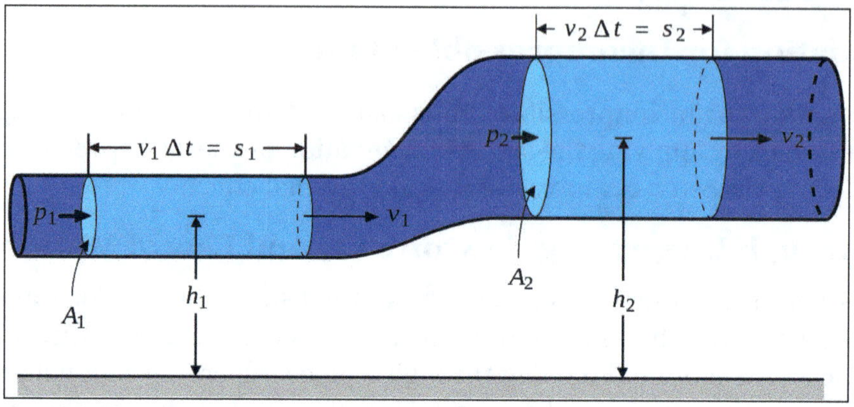

A streamtube of fluid moving to the right. Indicated are pressure, elevation, flow speed, distance (s), and cross-sectional area. Note that in this figure elevation is denoted as h, contrary to the text where it is given by z.

Derivation by using Conservation of Energy

Another way to derive Bernoulli's principle for an incompressible flow is by applying conservation of energy. In the form of the work-energy theorem, stating that

> *the change in the kinetic energy* E_{kin} *of the system equals the net work* W *done on the system*;
>
> $W = \Delta E_{\text{kin}}.$

Therefore,

> *the work done by the forces in the fluid = increase in kinetic energy.*

The system consists of the volume of fluid, initially between the cross-sections A_1 and A_2. In the time interval Δt fluid elements initially at the inflow cross-section A_1 move over a distance $s_1 = v_1 \Delta t$, while at the outflow cross-section the fluid moves away from cross-section A_2 over a distance $s_2 = v_2 \Delta t$. The displaced fluid volumes at the inflow and outflow are respectively $A_1 s_1$ and $A_2 s_2$. The associated displaced fluid masses are – when ρ is the fluid's mass density – equal to density times volume, so $\rho A_1 s_1$ and $\rho A_2 s_2$. By mass conservation, these two masses displaced in the time interval Δt have to be equal, and this displaced mass is denoted by Δm:

$$\rho A_1 s_1 = \rho A_1 v_1 \Delta t = \Delta m,$$

$$\rho A_2 s_2 = \rho A_2 v_2 \Delta t = \Delta m.$$

The work done by the forces consists of two parts:

- The *work done by the pressure* acting on the areas A_1 and A_2.

$$W_{\text{pressure}} = F_{1,\text{pressure}} s_1 - F_{2,\text{pressure}} s_2 = p_1 A_1 s_1 - p_2 A_2 s_2 = \Delta m \frac{p_1}{\rho} - \Delta m \frac{p_2}{\rho}.$$

- The *work done by gravity*: the gravitational potential energy in the volume $A_1 s_1$ is lost, and at the outflow in the volume $A_2 s_2$ is gained. So, the change in gravitational potential energy $\Delta E_{\text{pot,gravity}}$ in the time interval Δt is

  $$\Delta E_{\text{pot,gravity}} = \Delta m g z_2 - \Delta m g z_1.$$

  Now, the work by the force of gravity is opposite to the change in potential energy, $W_{\text{gravity}} = -\Delta E_{\text{pot,gravity}}$: while the force of gravity is in the negative *z*-direction, the work—gravity force times change in elevation—will be negative for a positive elevation change $\Delta z = z_2 - z_1$, while the corresponding potential energy change is positive. So:

  $$W_{\text{gravity}} = -\Delta E_{\text{pot,gravity}} = \Delta m g z_1 - \Delta m g z_2.$$

And the total work done in this time interval Δt is,

$$W = W_{\text{pressure}} + W_{\text{gravity}}.$$

The *increase in kinetic energy* is,

$$\Delta E_{\text{kin}} = \tfrac{1}{2}\Delta m v_2^2 - \tfrac{1}{2}\Delta m v_1^2.$$

Putting these together, the work-kinetic energy theorem $W = \Delta E_{\text{kin}}$ gives:

$$\Delta m \frac{p_1}{\rho} - \Delta m \frac{p_2}{\rho} + \Delta m g z_1 - \Delta m g z_2 = \frac{1}{2}\Delta m v_2^2 - \frac{1}{2}\Delta m v_1^2$$

or

$$\frac{1}{2}\Delta m v_1^2 + \Delta m g z_1 + \Delta m \frac{p_1}{\rho} = \frac{1}{2}\Delta m v_2^2 + \Delta m g z_2 + \Delta m \frac{p_2}{\rho}.$$

After dividing by the mass $\Delta m = \rho A_1 v_1 \Delta t = \rho A_2 v_2 \Delta t$ the result is:

$$\frac{1}{2} v_1^2 + g z_1 + \frac{p_1}{\rho} = \frac{1}{2} v_2^2 + g z_2 + \frac{p_2}{\rho}$$

or, as stated in the first paragraph:

$$\frac{v^2}{2} + gz + \frac{p}{\rho} = C$$ (Eqn. 1), Which is also Equation (A)

Further division by g produces the following equation. Note that each term can be described in the length dimension (such as meters). This is the head equation derived from Bernoulli's principle:

$$\frac{v^2}{2g} + z + \frac{p}{\rho g} = C \quad \text{(Eqn. 2a)}$$

The middle term, z, represents the potential energy of the fluid due to its elevation with respect to a reference plane. Now, z is called the elevation head and given the designation $z_{elevation}$.

A free falling mass from an elevation $z > 0$ (in a vacuum) will reach a speed,

$$v = \sqrt{2gz},$$

when arriving at elevation $z = 0$. Or when we rearrange it as a *head*:

$$h_v = \frac{v^2}{2g}$$

The term $v^2 / (2g)$ is called the *velocity head*, expressed as a length measurement. It represents the internal energy of the fluid due to its motion. The hydrostatic pressure *p* is defined as

$$p = p_0 - \rho g z,$$

with p_0 some reference pressure, or when we rearrange it as a *head*:

$$\psi = \frac{p}{\rho g}.$$

The term $p / (\rho g)$ is also called the *pressure head*, expressed as a length measurement. It represents the internal energy of the fluid due to the pressure exerted on the container. When we combine the head due to the flow speed and the head due to static pressure with the elevation above a reference plane, we obtain a simple relationship useful for incompressible fluids using the velocity head, elevation head, and pressure head.

$$h_v + z_{elevation} + \psi = C \quad \text{(Eqn. 2b)}$$

If we were to multiply Eqn. 1 by the density of the fluid, we would get an equation with three pressure terms:

$$\frac{\rho v^2}{2} + \rho g z + p = C \quad \text{(Eqn. 3)}$$

We note that the pressure of the system is constant in this form of the Bernoulli Equation. If the static pressure of the system (the far right term) increases, and if the pressure due to elevation (the middle term) is constant, then we know that the dynamic pressure (the left term) must have decreased. In other words, if the speed of a fluid decreases and it is not due to an elevation difference, we know it must be due to an increase in the static pressure that is resisting the flow.

All three equations are merely simplified versions of an energy balance on a system.

Bernoulli Equation for Compressible Fluids

The derivation for compressible fluids is similar. Again, the derivation depends upon (1) conservation of mass, and (2) conservation of energy. Conservation of mass implies that in the above figure, in the interval of time Δt, the amount of mass passing through the boundary defined by the area A_1 is equal to the amount of mass passing outwards through the boundary defined by the area A_2:

$$0 = \Delta M_1 - \Delta M_2 = \rho_1 A_1 v_1 \Delta t - \rho_2 A_2 v_2 \Delta t \,.$$

Conservation of energy is applied in a similar manner: It is assumed that the change in energy of the volume of the streamtube bounded by A_1 and A_2 is due entirely to energy entering or leaving through one or the other of these two boundaries. Clearly, in a more complicated situation such as a fluid flow coupled with radiation, such conditions are not met. Nevertheless, assuming this to be the case and assuming the flow is steady so that the net change in the energy is zero,

$$0 = \Delta E_1 - \Delta E_2$$

where ΔE_1 and ΔE_2 are the energy entering through A_1 and leaving through A_2, respectively. The energy entering through A_1 is the sum of the kinetic energy entering, the energy entering in the form of potential gravitational energy of the fluid, the fluid thermodynamic internal energy per unit of mass (ε_1) entering, and the energy entering in the form of mechanical $p\, dV$ work:

$$\Delta E_1 = \left[\frac{1}{2}\rho_1 v_1^2 + \Psi_1 \rho_1 + \epsilon_1 \rho_1 + p_1\right] A_1 v_1 \Delta t$$

where $\Psi = gz$ is a force potential due to the Earth's gravity, g is acceleration due to gravity, and z is elevation above a reference plane. A similar expression for may easily be constructed. So now setting $0 = \Delta E_1 - \Delta E_2$:

$$0 = \left[\frac{1}{2}\rho_1 v_1^2 + \Psi_1 \rho_1 + \epsilon_1 \rho_1 + p_1\right] A_1 v_1 \Delta t - \left[\frac{1}{2}\rho_2 v_2^2 + \Psi_2 \rho_2 + \epsilon_2 \rho_2 + p_2\right] A_2 v_2 \Delta t$$

which can be rewritten as:

$$0 = \left[\frac{1}{2} v_1^2 + \Psi_1 + \epsilon_1 + \frac{p_1}{\rho_1}\right] \rho_1 A_1 v_1 \Delta t - \left[\frac{1}{2} v_2^2 + \Psi_2 + \epsilon_2 + \frac{p_2}{\rho_2}\right] \rho_2 A_2 v_2 \Delta t$$

Now, using the previously-obtained result from conservation of mass, this may be simplified to obtain,

$$\frac{1}{2} v^2 + \Psi + \epsilon + \frac{p}{\rho} = \text{constant} \equiv b$$

which is the Bernoulli equation for compressible flow.

An equivalent expression can be written in terms of fluid enthalphy (h):

$$\frac{1}{2} v^2 + \Psi + h = \text{constant} \equiv b$$

Applications

Condensation visible over the upper surface of an Airbus A340 wing caused by the fall in temperature accompanying the fall in pressure.

In modern everyday life there are many observations that can be successfully explained by application of Bernoulli's principle, even though no real fluid is entirely inviscid and a small viscosity often has a large effect on the flow:

- Bernoulli's principle can be used to calculate the lift force on an airfoil, if the behaviour of the fluid flow in the vicinity of the foil is known. For example, if the air flowing past the top surface of an aircraft wing is moving faster than the air flowing past the bottom surface, then Bernoulli's principle implies that the pressure on the surfaces of the wing will be lower above than below. This pressure difference results in an upwards lifting force. Whenever the distribution of speed past the top and bottom surfaces of a wing is known, the lift forces can be calculated (to a good approximation) using Bernoulli's equations – established by Bernoulli over a century before the first man-made wings were used for the purpose of flight. Bernoulli's principle does not explain why the air flows faster past the top of the wing and slower past the underside.
- The carburettor used in many reciprocating engines contains a venturi to create a region of low pressure to draw fuel into the carburettor and mix it thoroughly with the incoming air. The low pressure in the throat of a venturi can be explained by Bernoulli's principle; in the narrow throat, the air is moving at its fastest speed and therefore it is at its lowest pressure.
- An injector on a steam locomotive (or static boiler).
- The pitot tube and static port on an aircraft are used to determine the airspeed of the aircraft. These two devices are connected to the airspeed indicator, which determines the dynamic pressure of the airflow past the aircraft. Dynamic pressure is the difference between stagnation pressure and static pressure. Bernoulli's principle is used to calibrate the airspeed indicator so that it displays the indicated airspeed appropriate to the dynamic pressure.

- The flow speed of a fluid can be measured using a device such as a Venturi meter or an orifice plate, which can be placed into a pipeline to reduce the diameter of the flow. For a horizontal device, the continuity equation shows that for an incompressible fluid, the reduction in diameter will cause an increase in the fluid flow speed. Subsequently Bernoulli's principle then shows that there must be a decrease in the pressure in the reduced diameter region. This phenomenon is known as the Venturi effect.
- The maximum possible drain rate for a tank with a hole or tap at the base can be calculated directly from Bernoulli's equation, and is found to be proportional to the square root of the height of the fluid in the tank. This is Torricelli's law, showing that Torricelli's law is compatible with Bernoulli's principle. Viscosity lowers this drain rate. This is reflected in the discharge coefficient, which is a function of the Reynolds number and the shape of the orifice.
- The Bernoulli grip relies on this principle to create a non-contact adhesive force between a surface and the gripper.

Misunderstandings About the Generation of Lift

Many explanations for the generation of lift (on airfoils, propeller blades, etc.) can be found; some of these explanations can be misleading, and some are false, in particular the idea that air particles flowing above and below a cambered wing should reach simultaneously the trailing edge. This has been a source of heated discussion over the years. In particular, there has been debate about whether lift is best explained by Bernoulli's principle or Newton's laws of motion. Modern writings agree that both Bernoulli's principle and Newton's laws are relevant and either can be used to correctly describe lift.

Several of these explanations use the Bernoulli principle to connect the flow kinematics to the flow-induced pressures. In cases of incorrect (or partially correct) explanations relying on the Bernoulli principle, the errors generally occur in the assumptions on the flow kinematics and how these are produced. It is not the Bernoulli principle itself that is questioned because this principle is well established (the airflow above the wing *is* faster, the question is *why* it is faster).

Misapplications of Bernoulli's Principle in Common Classroom Demonstrations

There are several common classroom demonstrations that are sometimes incorrectly explained using Bernoulli's principle. One involves holding a piece of paper horizontally so that it droops downward and then blowing over the top of it. As the demonstrator blows over the paper, the paper rises. It is then asserted that this is because "faster moving air has lower pressure".

One problem with this explanation can be seen by blowing along the bottom of the paper: were the deflection due simply to faster moving air one would expect the paper to deflect downward, but the paper deflects upward regardless of whether the faster moving air is on the top or the bottom. Another problem is that when the air leaves the demonstrator's mouth it has the *same* pressure as the surrounding air; the air does not have lower pressure just because it is moving; in the demonstration, the static pressure of the air leaving the demonstrator's mouth is *equal* to the pressure of the surrounding air. A third problem is that it is false to make a connection between the flow on the two sides of the paper using Bernoulli's equation since the air above and below are *different* flow fields and Bernoulli's principle only applies within a flow field.

As the wording of the principle can change its implications, stating the principle correctly is important. What Bernoulli's principle actually says is that within a flow of constant energy, when fluid flows through a region of lower pressure it speeds up and vice versa. Thus, Bernoulli's principle concerns itself with *changes* in speed and *changes* in pressure *within* a flow field. It cannot be used to compare different flow fields.

A correct explanation of why the paper rises would observe that the plume follows the curve of the paper and that a curved streamline will develop a pressure gradient perpendicular to the direction of flow, with the lower pressure on the inside of the curve. Bernoulli's principle predicts that the decrease in pressure is associated with an increase in speed, i.e. that as the air passes over the paper it speeds up and moves faster than it was moving when it left the demonstrator's mouth. But this is not apparent from the demonstration.

Other common classroom demonstrations, such as blowing between two suspended spheres, inflating a large bag, or suspending a ball in an airstream are sometimes explained in a similarly misleading manner by saying "faster moving air has lower pressure".

Physical Law

A physical law or scientific law is a theoretical statement "inferred from particular facts, applicable to a defined group or class of phenomena, and expressible by the statement that a particular phenomenon always occurs if certain conditions be present." Physical laws are typically conclusions based on repeated scientific experiments and observations over many years and which have become accepted universally within the scientific community. The production of a summary description of our environment in the form of such laws is a fundamental aim of science. These terms are not used the same way by all authors.

The distinction between natural law in the political-legal sense and law of nature or physical law in the scientific sense is a modern one, both concepts being equally derived from *physis*, the Greek word (translated into Latin as *natura*) for *nature*.

Description

Several general properties of physical laws have been identified. Physical laws are:

- True, at least within their regime of validity. By definition, there have never been repeatable contradicting observations.
- Universal. They appear to apply everywhere in the universe.
- Simple. They are typically expressed in terms of a single mathematical equation.
- Absolute. Nothing in the universe appears to affect them.
- Stable. Unchanged since first discovered (although they may have been shown to be approximations of more accurate laws.

- Omnipotent. Everything in the universe apparently must comply with them (according to observations).
- Generally conservative of quantity.
- Often expressions of existing homogeneities (symmetries) of space and time.
- Typically theoretically reversible in time (if non-quantum), although time itself is irreversible.

Examples

Some of the more famous laws of nature are found in Isaac Newton's theories of (now) classical mechanics, presented in his *Philosophiae Naturalis Principia Mathematica*, and in Albert Einstein's theory of relativity. Other examples of laws of nature include Boyle's law of gases, conservation laws, the four laws of thermodynamics, etc.

Laws as Definitions

Many scientific laws are couched in mathematical terms (e.g. Newton's Second law $F = {}^{dp}/_{dt}$, or the uncertainty principle, or the principle of least action, or causality). While these scientific laws explain what our senses perceive, they are still empirical, and so are not "mathematical" laws. (Mathematical laws can be proved purely by mathematics and not by scientific experiment.)

Laws Being Consequences of Mathematical Symmetries

Other laws reflect mathematical symmetries found in Nature (say, Pauli exclusion principle reflects identity of electrons, conservation laws reflect homogeneity of space, time, Lorentz transformations reflect rotational symmetry of space–time). Laws are constantly being checked experimentally to higher and higher degrees of precision. This is one of the main goals of science. Just because laws have never been observed to be violated does not preclude testing them at increased accuracy or in new kinds of conditions to confirm whether they continue to hold, or whether they break, and what can be discovered in the process. It is always possible for laws to be invalidated or proven to have limitations, by repeatable experimental evidence, should any be observed.

Well-established laws have indeed been invalidated in some special cases, but the new formulations created to explain the discrepancies can be said to generalize upon, rather than overthrow, the originals. That is, the invalidated laws have been found to be only close approximations, to which other terms or factors must be added to cover previously unaccounted-for conditions, e.g. very large or very small scales of time or space, enormous speeds or masses, etc. Thus, rather than unchanging knowledge, physical laws are better viewed as a series of improving and more precise generalizations.

Laws as Approximations

Some laws are only approximations of other more general laws, and are good approximations with a restricted domain of applicability. For example, Newtonian dynamics (which is based on Galilean transformations) is the low speed limit of special relativity (since the Galilean transformation is the low-speed approximation to the Lorentz transformation). Similarly, the Newtonian gravitation law is a low-mass approximation of general relativity, and Coulomb's law is an approximation to Quantum Electrodynamics at large distances (compared to the range of weak interactions). In such cases it is common to use the simpler, approximate versions of the laws, instead of the more accurate general laws.

Physical Laws Derived from Symmetry Principles

Many fundamental physical laws are mathematical consequences of various symmetries of space, time, or other aspects of nature. Specifically, Noether's theorem connects some conservation laws to certain symmetries. For example, conservation of energy is a consequence of the shift symmetry of time (no moment of time is different from any other), while conservation of momentum is a consequence of the symmetry (homogeneity) of space (no place in space is special, or different than any other). The indistinguishability of all particles of each fundamental type (say, electrons, or photons) results in the Dirac and Bose quantum statistics which in turn result in the Pauli exclusion principle for fermions and in Bose–Einstein condensation for bosons. The rotational symmetry between time and space coordinate axes (when one is taken as imaginary, another as real) results in Lorentz transformations which in turn result in special relativity theory. Symmetry between inertial and gravitational mass results in general relativity.

The inverse square law of interactions mediated by massless bosons is the mathematical consequence of the 3-dimensionality of space.

One strategy in the search for the most fundamental laws of nature is to search for the most general mathematical symmetry group that can be applied to the fundamental interactions.

History

Compared to pre-modern accounts of causality, laws of nature fill the role played by divine causality on the one hand, and accounts such as Plato's theory of forms on the other.

The observation that there are underlying regularities in nature dates from prehistoric times, since the recognition of cause-and-effect relationships is an implicit recognition that there are laws of nature. The recognition of such regularities as independent scientific laws *per se*, though, was limited by their entanglement in animism, and by the attribution of many effects that do not have readily obvious causes—such as meteorological, astronomical and biological phenomena—to the actions of various gods, spirits, supernatural beings, etc. Observation and speculation about nature were intimately bound up with metaphysics and morality.

In Europe, systematic theorizing about nature (*physis*) began with the early Greek philosophers and scientists and continued into the Hellenistic and Roman imperial periods, during which times the intellectual influence of Roman law increasingly became paramount.

The formula "law of nature" first appears as "a live metaphor" favored by Latin poets Lucretius, Virgil, Ovid, Manilius, in time gaining a firm theoretical presence in the prose treatises of Seneca and Pliny. Why this Roman origin? According to [historian and classicist Daryn] Lehoux's persuasive narrative, the idea was made possible by the pivotal role of codified law and forensic argument in Roman life and culture.

For the Romans . . . the place par excellence where ethics, law, nature, religion and politics overlap is the law court. When we read Seneca's *Natural Questions*, and watch again and again just how he applies standards of evidence, witness evaluation, argument and proof, we can recognize that we are reading one of the great Roman rhetoricians of the age, thoroughly immersed in forensic method. And not Seneca alone. Legal models of scientific judgment turn up all over the place, and for example prove equally integral to Ptolemy's approach to verification, where the mind is assigned the role of magistrate, the senses that of disclosure of evidence, and dialectical reason that of the law itself.

The precise formulation of what are now recognized as modern and valid statements of the laws of nature dates from the 17th century in Europe, with the beginning of accurate experimentation and development of advanced forms of mathematics. During this period, natural philosophers such as Isaac Newton were influenced by a religious view which held that God had instituted absolute, universal and immutable physical laws. The modern scientific method which took shape at this time (with Francis Bacon and Galileo) aimed at total separation of science from theology, with minimal speculation about metaphysics and ethics. Natural law in the political sense, conceived as universal (i.e., divorced from sectarian religion and accidents of place), was also elaborated in this period (by Grotius, Spinoza, and Hobbes, to name a few).

Other Fields

Some mathematical theorems and axioms are referred to as laws because they provide logical foundation to empirical laws.

Examples of other observed phenomena sometimes described as laws include the Titius-Bode law of planetary positions, Zipf's law of linguistics, Moore's law of technological growth. Many of these laws fall within the scope of uncomfortable science. Other laws are pragmatic and observational, such as the law of unintended consequences. By analogy, principles in other fields of study are sometimes loosely referred to as "laws". These include Occam's razor as a principle of philosophy and the Pareto principle of economics.

Conservation Law

In physics, a conservation law states that a particular measurable property of an isolated physical system does not change as the system evolves over time. Exact conservation laws include conservation of energy, conservation of linear momentum, conservation of angular momentum, and conservation of electric charge. There are also many approximate conservation laws, which apply to such quantities as mass, parity, lepton number, baryon number, strangeness, hypercharge, etc. These quantities are conserved in certain classes of physics processes, but not in all.

A local conservation law is usually expressed mathematically as a continuity equation, a partial differential equation which gives a relation between the amount of the quantity and the "transport" of that quantity. It states that the amount of the conserved quantity at a point or within a volume can only change by the amount of the quantity which flows in or out of the volume.

From Noether's theorem, each conservation law is associated with a symmetry in the underlying physics.

Conservation Laws as Fundamental Laws of Nature

Conservation laws are fundamental to our understanding of the physical world, in that they describe which processes can or cannot occur in nature. For example, the conservation law of energy states that the total quantity of energy in an isolated system does not change, though it may change form. In general, the total quantity of the property governed by that law remains unchanged during physical processes. With respect to classical physics, conservation laws include conservation of energy, mass (or matter), linear momentum, angular momentum, and electric charge. With respect to particle physics, particles cannot be created or destroyed except in pairs, where one is ordinary and the other is an antiparticle. With respect to symmetries and invariance principles, three special conservation laws have been described, associated with inversion or reversal of space, time, and charge.

Conservation laws are considered to be fundamental laws of nature, with broad application in physics, as well as in other fields such as chemistry, biology, geology, and engineering.

Most conservation laws are exact, or absolute, in the sense that they apply to all possible processes. Some conservation laws are partial, in that they hold for some processes but not for others.

One particularly important result concerning conservation laws is Noether's theorem, which states that there is a one-to-one correspondence between each one of them and a differentiable symmetry in the system. For example, the conservation of energy follows from the time-invariance of physical systems, and the fact that physical systems behave the same regardless of how they are oriented in space gives rise to the conservation of angular momentum.

Exact Laws

A partial listing of physical conservation equations due to symmetry that are said to be exact laws, or more precisely *have never been proven to be violated:*

<table>
<tr><th>Conservation Law</th><th colspan="2">Respective Noether symmetry invariance</th><th colspan="2">Number of dimensions</th></tr>
<tr><td>Conservation of mass-energy</td><td>Time invariance</td><td rowspan="4">Lorentz invariance symmetry</td><td>1</td><td>translation about time axis</td></tr>
<tr><td>Conservation of linear momentum</td><td>Translation symmetry</td><td>3</td><td>translation about x,y,z position</td></tr>
<tr><td>Conservation of angular momentum</td><td>Rotation invariance</td><td>3</td><td>rotation about x,y,z axes</td></tr>
<tr><td>CPT symmetry (combining charge, parity and time conjugation)</td><td>Lorentz invariance</td><td>1+1+1</td><td>(charge inversion q→-q) + (position inversion r→-r) + (time inversion t→-t)</td></tr>
<tr><td>Conservation of electric charge</td><td colspan="2">Gauge invariance</td><td>1→4</td><td>scalar field (1D) in 4D spacetime (x,y,z + time evolution)</td></tr>
</table>

Conservation of color charge	SU(3) Gauge invariance	3	r,g,b
Conservation of weak isospin	$SU(2)_L$ Gauge invariance	1	weak charge
Conservation of probability	Probability invariance	1→4	total probability always=1 in whole x,y,z space, during time evolution

Approximate Laws

There are also approximate conservation laws. These are approximately true in particular situations, such as low speeds, short time scales, or certain interactions.

- Conservation of rest mass.
- Conservation of baryon number.
- Conservation of lepton number (In the Standard Model).
- Conservation of flavor (violated by the weak interaction).
- Conservation of parity.
- Invariance under charge conjugation.
- Invariance under time reversal.
- CP symmetry, the combination of charge and parity conjugation (equivalent to time reversal if CPT holds).

Differential Forms

In continuum mechanics, the most general form of an exact conservation law is given by a continuity equation. For example, conservation of electric charge q is

$$\frac{\partial \rho}{\partial t} = -\nabla \cdot \mathbf{j}$$

where $\nabla\cdot$ is the divergence operator, ρ is the density of q (amount per unit volume), j is the flux of q (amount crossing a unit area in unit time), and t is time.

If we assume that the motion u of the charge is a continuous function of position and time, then

$$\mathbf{j} = \rho\mathbf{u}$$

$$\frac{\partial \rho}{\partial t} = -\nabla \cdot (\rho\mathbf{u}).$$

In one space dimension this can be put into the form of a homogeneous first-order quasilinear hyperbolic equation:

$$y_t + A(y)y_x = 0$$

where the dependent variable y is called the *density* of a *conserved quantity*, and $A(y)$ is called the *current jacobian*, and the subscript notation for partial derivatives has been employed. The more general inhomogeneous case:

$$y_t + A(y)y_x = s$$

is not a conservation equation but the general kind of balance equation describing a dissipative system. The dependent variable *y* is called a *nonconserved quantity*, and the inhomogeneous term *s(y,x,t)* is the-*source*, or dissipation. For example, balance equations of this kind are the momentum and energy Navier-Stokes equations, or the entropy balance for a general isolated system.

In the one-dimensional space a conservation equation is a first-order quasilinear hyperbolic equation that can be put into the *advection* form:

$$y_t + a(y)y_x = 0$$

where the dependent variable *y(x,t)* is called the density of the *conserved* (scalar) quantity (c.q.(d.) = conserved quantity (density)), and *a(y)* is called the current coefficient, usually corresponding to the partial derivative in the conserved quantity of a current density (c.d.) of the conserved quantity *j(y)*:

$$a(y) = j_y(y)$$

In this case since the chain rule applies:

$$j_x = j_y(y)y_x = a(y)y_x$$

the conservation equation can be put into the current density form:

$$y_t + j_x(y) = 0$$

In a space with more than one dimension the former definition can be extended to an equation that can be put into the form:

$$y_t + \mathbf{a}(y) \cdot \nabla y = 0$$

where the *conserved quantity* is *y(r,t)*, $\cdot$ denotes the scalar product, ∇ is the nabla operator, here indicating a gradient, and *a(y)* is a vector of current coefficients, analogously corresponding to the divergence of a vector c.d. associated to the c.q. j(y):

$$y_t + \nabla \cdot \mathbf{j}(y) = 0$$

This is the case for the continuity equation:

$$\rho_t + \nabla \cdot (\rho \mathbf{u}) = 0$$

Here the conserved quantity is the mass, with density ρ(r,t) and current density ρu, identical to the momentum density, while u(r,t) is the flow velocity.

In the general case a conservation equation can be also a system of this kind of equations (a vector equation) in the form:

$$\mathbf{y}_t + \mathbf{A}(\mathbf{y}) \cdot \nabla \mathbf{y} = \mathbf{0}$$

where y is called the *conserved* (vector) quantity, ∇ y is its gradient, 0 is the zero vector, and A(y) is called the Jacobian of the current density. In fact as in the former scalar case, also in the vector case A(y) usually corresponding to the Jacobian of a current density matrix J(y),

$$\mathbf{A}(\mathbf{y}) = \mathbf{J}_{\mathbf{y}}(\mathbf{y})$$

and the conservation equation can be put into the form:

$$\mathbf{y}_t + \nabla \cdot \mathbf{J}(\mathbf{y}) = \mathbf{0}$$

For example, this the case for Euler equations (fluid dynamics). In the simple incompressible case they are:

$$\nabla \cdot \mathbf{u} = 0$$

$$\frac{\partial \mathbf{u}}{\partial t} + \mathbf{u} \cdot \nabla \mathbf{u} + \nabla s = \mathbf{0},$$

where:

- u is the flow velocity vector, with components in a N-dimensional space $u_1, u_2 \dots u_N$,
- s is the specific pressure (pressure per unit density) giving the source term,

It can be shown that the conserved (vector) quantity and the c.d. matrix for these equations are respectively:

$$\mathbf{y} = \begin{pmatrix} 1 \\ \mathbf{u} \end{pmatrix}; \qquad \mathbf{J} = \begin{pmatrix} \mathbf{u} \\ \mathbf{u} \otimes \mathbf{u} + s\mathbf{I} \end{pmatrix};$$

where $\otimes$ denotes the outer product.

Integral and Weak Forms

Conservation equations can be also expressed in integral form: the advantage of the latter is substantially that it requires less smoothness of the solution, which paves the way to weak form, extending the class of admissible solutions to include discontinuous solutions. By integrating in any space-time domain the current density form in 1-D space:

$$y_t + j_x(y) = 0$$

and by using Green's theorem, the integral form is:

$$\int_{-\infty}^{\infty} y dx + \int_0^{\infty} j(y) dt = 0$$

In a similar fashion, for the scalar multidimensional space, the integral form is:

$$\oint [y d^N r + j(y) dt] = 0$$

where the line integration is performed along the boundary of the domain, in an anticlock-wise manner.

Moreover, by defining a test function φ(r,t) continuously differentiable both in time and space with compact support, the weak form can be obtained pivoting on the initial condition. In 1-D space it is:

$$\int_{0}^{\infty}\int_{-\infty}^{\infty} \phi_t y + \phi_x j(y) dx dt = -\int_{-\infty}^{\infty} \phi(x,0) y(x,0) dx$$

Note that in the weak form all the partial derivatives of the density and current density have been passed on to the test function, which with the former hypothesis is sufficiently smooth to admit these derivatives.

Inverse-square Law

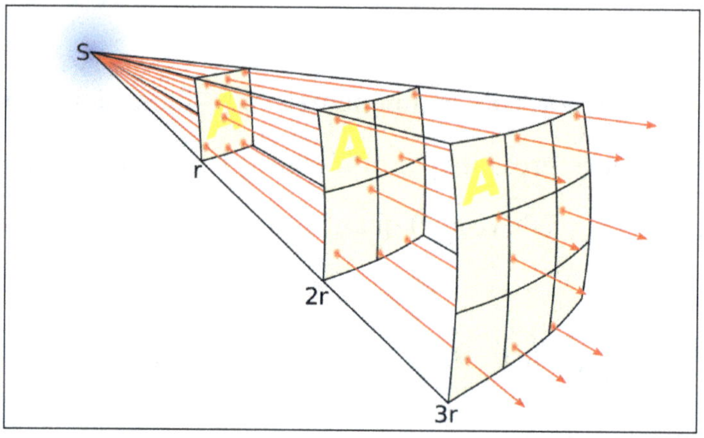

S represents the light source, while r represents the measured points. The lines represent the flux emanating from the source. The total number of flux lines depends on the strength of the source and is constant with increasing distance. A greater density of flux lines (lines per unit area) means a stronger field. The density of flux lines is inversely proportional to the square of the distance from the source because the surface area of a sphere increases with the square of the radius. Thus the strength of the field is inversely proportional to the square of the distance from the source.

In physics, an inverse-square law is any physical law stating that a specified physical quantity or intensity is inversely proportional to the square of the distance from the source of that physical quantity. The fundamental cause for this can be understood as geometric dilution corresponding to point-source radiation into three-dimensional space. Mathematically formulated:

$$\text{intensity} \propto \frac{1}{\text{distance}^2}$$

It can also be mathematically expressed as:

$$\frac{\text{intensity}_1}{\text{intensity}_2} = \frac{\text{distance}_2{}^2}{\text{distance}_1{}^2}$$

or as the formulation of a constant quantity:

$$\text{Intensity}_1 \times \text{distance}_1{}^2 = \text{Intensity}_2 \times \text{distance}_2{}^2$$

The divergence of a vector field which is the resultant of radial inverse-square law fields with respect to one or more sources is everywhere proportional to the strength of the local sources, and hence zero outside sources. Newton's law of universal gravitation follows an inverse-square law, as do the effects of electric, magnetic, light, sound, and radiation phenomena.

Radar energy expands during both the signal transmission and also on the reflected return, so the inverse square for both paths means that the radar will receive energy according to $1/4r^2$ power.

In order to prevent dilution of energy while propagating a signal, certain methods can be used such as a waveguide, which acts like a canal does for water, or how a gun barrel restricts hot gas expansion to one dimension in order to prevent loss of energy transfer to a bullet.

Justification

The inverse-square law generally applies when some force, energy, or other conserved quantity is evenly radiated outward from a point source in three-dimensional space. Since the surface area of a sphere (which is $4\pi r^2$) is proportional to the square of the radius, as the emitted radiation gets farther from the source, it is spread out over an area that is increasing in proportion to the square of the distance from the source. Hence, the intensity of radiation passing through any unit area (directly facing the point source) is inversely proportional to the square of the distance from the point source. Gauss's law is similarly applicable, and can be used with any physical quantity that acts in accord to the inverse-square relationship.

Occurrences

Gravitation

Gravitation is the attraction of two objects with mass. Newton's law states:

> *The gravitational attraction force between two point masses is directly proportional to the product of their masses and inversely proportional to the square of their separation distance. The force is always attractive and acts along the line joining them.*

If the distribution of matter in each body is spherically symmetric, then the objects can be treated as point masses without approximation, as shown in the shell theorem. Otherwise, if we want to calculate the attraction between massive bodies, we need to add all the point-point attraction forces vectorially and the net attraction might not be exact inverse square. However, if the separation between the massive bodies is much larger compared to their sizes, then to a good approximation, it is reasonable to treat the masses as point mass while calculating the gravitational force.

As the law of gravitation, this law was suggested in 1645 by Ismael Bullialdus. But Bullialdus did not accept Kepler's second and third laws, nor did he appreciate Christiaan Huygens's solution for circular motion (motion in a straight line pulled aside by the central force). Indeed, Bullialdus maintained the sun's force was attractive at aphelion and repulsive at perihelion. Robert Hooke and Giovanni Alfonso Borelli both expounded gravitation in 1666 as an attractive force (Hooke's lecture "On gravity" at the Royal Society, London, on 21 March; Borelli's "Theory of the Planets", published later in 1666). Hooke's 1670 Gresham lecture explained that gravitation applied to "all celestiall bodys" and added the principles that the gravitating power decreases with distance and that in the absence of any such power bodies move in straight lines. By 1679, Hooke thought gravitation had inverse square dependence and communicated this in a letter to Isaac Newton. Hooke remained bitter about Newton claiming the invention of this principle, even though Newton's *Principia* acknowledged that Hooke, along with Wren and Halley, had separately appreciated the inverse square law in the solar system, as well as giving some credit to Bullialdus.

Electrostatics

The force of attraction or repulsion between two electrically charged particles, in addition to being directly proportional to the product of the electric charges, is inversely proportional to the square of the distance between them; this is known as Coulomb's law. The deviation of the exponent from 2 is less than one part in 10^{15}.

Light and Other Electromagnetic Radiation

The intensity (or illuminance or irradiance) of light or other linear waves radiating from a point source (energy per unit of area perpendicular to the source) is inversely proportional to the square of the distance from the source; so an object (of the same size) twice as far away, receives only one-quarter the energy (in the same time period).

More generally, the irradiance, *i.e.*, the intensity (or power per unit area in the direction of propagation), of a spherical wavefront varies inversely with the square of the distance from the source (assuming there are no losses caused by absorption or scattering).

For example, the intensity of radiation from the Sun is 9126 watts per square meter at the distance of Mercury (0.387 AU); but only 1367 watts per square meter at the distance of Earth (1 AU)—an approximate threefold increase in distance results in an approximate ninefold decrease in intensity of radiation.

For non-isotropic radiators such as parabolic antennas, headlights, and lasers, the effective origin is located far behind the beam aperture. If you are close to the origin, you don't have to go far to double the radius, so the signal drops quickly. When you are far from the origin and still have a strong signal, like with a laser, you have to travel very far to double the radius and reduce the signal. This means you have a stronger signal or have antenna gain in the direction of the narrow beam relative to a wide beam in all directions of an isotropic antenna.

In photography and stage lighting, the inverse-square law is used to determine the "fall off" or the difference in illumination on a subject as it moves closer to or further from the light source. For quick approximations, it is enough to remember that doubling the distance reduces illumination to one quarter; or similarly, to halve the illumination increase the distance by a factor of 1.4 (the square root of 2), and to double illumination, reduce the distance to 0.7 (square root of 1/2). When the illuminant is not a point source, the inverse square rule is often still a useful approximation; when the size of the light source is less than one-fifth of the distance to the subject, the calculation error is less than 1%.

The fractional reduction in electromagnetic fluence (Φ) for indirectly ionizing radiation with increasing distance from a point source can be calculated using the inverse-square law. Since emissions from a point source have radial directions, they intercept at a perpendicular incidence.

The area of such a shell is $4\pi r^2$ where r is the radial distance from the center. The law is particularly important in diagnostic radiography and radiotherapy treatment planning, though this proportionality does not hold in practical situations unless source dimensions are much smaller than the distance. As stated in fourier theory of heat "as the point source is magnification by distances , its radiation is dilute proportional to the sin of the angle, of the increasing circumference arc from the point of origin"

Example

Let the total power radiated from a point source, for example, an omnidirectional isotropic radiator, be P. At large distances from the source (compared to the size of the source), this power is distributed over larger and larger spherical surfaces as the distance from the source increases. Since the surface area of a sphere of radius r is $A = 4\pi r^2$, then intensity I (power per unit area) of radiation at distance r is

$$I = \frac{P}{A} = \frac{P}{4\pi r^2}.$$

The energy or intensity decreases (divided by 4) as the distance r is doubled; measured in dB it would decrease by 6.02 dB per doubling of distance.

Sound in a Gas

In acoustics, the sound pressure of a spherical wavefront radiating from a point source decreases by 50% as the distance r is doubled; measured in dB, the decrease is still 6.02 dB, since dB represents an intensity ratio. The pressure ratio (as opposed to power ratio) is not inverse-square, but is inverse-proportional (inverse distance law):

$$\mathrm{p} \propto \frac{1}{\mathrm{r}}$$

The same is true for the component of particle velocity v that is in-phase with the instantaneous sound pressure p:

$$v \propto \frac{1}{r}$$

In the near field is a quadrature component of the particle velocity that is 90° out of phase with the sound pressure and does not contribute to the time-averaged energy or the intensity of the sound. The sound intensity is the product of the RMS sound pressure and the in-phase component of the RMS particle velocity, both of which are inverse-proportional. Accordingly, the intensity follows an inverse-square behaviour:

$$\mathrm{I} = \mathrm{pv} \propto \frac{1}{\mathrm{r}^2}.$$

Field Theory Interpretation

For an irrotational vector field in three-dimensional space, the inverse-square law corresponds to the property that the divergence is zero outside the source. This can be generalized to higher dimensions. Generally, for an irrotational vector field in n-dimensional Euclidean space, the intensity "I" of the vector field falls off with the distance "r" following the inverse $(n-1)^{\text{th}}$ power law

$$I \propto \frac{1}{r^{n-1}},$$

given that the space outside the source is divergence free.

History

John Dumbleton of the 14th-century Oxford Calculators, was one of the first to express functional relationships in graphical form. He gave a proof of the mean speed theorem stating that "the latitude of a uniformly difform movement corresponds to the degree of the midpoint" and used this method to study the quantitative decrease in intensity of illumination in his *Summa logicæ et philosophiæ naturalis* (ca. 1349), stating that it was not linearly proportional to the distance, but was unable to expose the Inverse-square law.

In proposition 9 of Book 1 in his book *Ad Vitellionem paralipomena, quibus astronomiae pars optica traditur* (1604), the astronomer Johannes Kepler argued that the spreading of light from a point source obeys an inverse square law:

Original: *Sicut se habent sphaericae superificies, quibus origo lucis pro centro est, amplior ad angustiorem: ita se habet fortitudo seu densitas lucis radiorum in angustiori, ad illamin in laxiori sphaerica, hoc est, conversim. Nam per 6. 7. tantundem lucis est in angustiori sphaerica superficie, quantum in fusiore, tanto ergo illie stipatior & densior quam hic.*

Translation: Just as [the ratio of] spherical surfaces, for which the source of light is the center, [is] from the wider to the narrower, so the density or fortitude of the rays of light in the narrower [space], towards the more spacious spherical surfaces, that is, inversely. For according to [propositions] 6 & 7, there is as much light in the narrower spherical surface, as in the wider, thus it is as much more compressed and dense here than there.

In 1645 in his book *Astronomia Philolaica* ..., the French astronomer Ismaël Bullialdus (1605 – 1694) refuted Johannes Kepler's suggestion that "gravity" weakens as the inverse of the distance; instead, Bullialdus argued, "gravity" weakens as the inverse square of the distance:

Original: *Virtus autem illa, qua Sol prehendit seu harpagat planetas, corporalis quae ipsi pro manibus est, lineis rectis in omnem mundi amplitudinem emissa quasi species solis cum illius corpore rotatur: cum ergo sit corporalis imminuitur, & extenuatur in maiori spatio & intervallo, ratio autem huius imminutionis eadem est, ac luminus, in ratione nempe dupla intervallorum, sed eversa.*

Translation: As for the power by which the Sun seizes or holds the planets, and which, being corporeal, functions in the manner of hands, it is emitted in straight lines throughout the whole extent of the world, and like the species of the Sun, it turns with the body of the Sun; now, seeing that it is corporeal, it becomes weaker and attenuated at a greater distance or interval, and the ratio of its decrease in strength is the same as in the case of light, namely, the duplicate proportion, but inversely, of the distances [that is, $1/d^2$].

In England, the Anglican bishop Seth Ward (1617 – 1689) publicized the ideas of Bullialdus in his critique *In Ismaelis Bullialdi astronomiae philolaicae fundamenta inquisitio brevis*

(1653) and publicized the planetary astronomy of Kepler in his book *Astronomia geometrica* (1656).

In 1663–1664, the English scientist Robert Hooke was writing his book *Micrographia* (1666) in which he discussed, among other things, the relation between the height of the atmosphere and the barometric pressure at the surface. Since the atmosphere surrounds the earth, which itself is a sphere, the volume of atmosphere bearing on any unit area of the earth's surface is a truncated cone (which extends from the earth's center to the vacuum of space; obviously only the section of the cone from the earth's surface to space bears on the earth's surface). Although the volume of a cone is proportional to the cube of its height, Hooke argued that the air's pressure at the earth's surface is instead proportional to the height of the atmosphere because gravity diminishes with altitude. Although Hooke did not explicitly state so, the relation that he proposed would be true only if gravity decreases as the inverse square of the distance from the earth's center.

Stokes' Law

In 1851, George Gabriel Stokes derived an expression, now known as Stokes' law, for the frictional force – also called drag force – exerted on spherical objects with very small Reynolds numbers (i.e. very small particles) in a viscous fluid. Stokes' law is derived by solving the Stokes flow limit for small Reynolds numbers of the Navier–Stokes equations.

Statement of the Law

The force of viscosity on a small sphere moving through a viscous fluid is given by:

where:

- F_d is the frictional force – known as Stokes' drag – acting on the interface between the fluid and the particle.
- η is the dynamic viscosity (Some authors use the symbol μ).
- r is the radius of the spherical object.
- v is the flow velocity relative to the object.

In SI units, F_d is given in Newtons, μ in Pa·s, R in meters, and V in m/s.

Stokes' law makes the following assumptions for the behavior of a particle in a fluid:

- Laminar Flow.
- Spherical particles.
- Homogeneous (uniform in composition) material.
- Smooth surfaces.

- Particles do not interfere with each other.

Note that for molecules Stokes' law is used to define their Stokes radius.

The CGS unit of kinematic viscosity was named "stokes" after his work.

Applications

Stokes' law is the basis of the falling-sphere viscometer, in which the fluid is stationary in a vertical glass tube. A sphere of known size and density is allowed to descend through the liquid. If correctly selected, it reaches terminal velocity, which can be measured by the time it takes to pass two marks on the tube. Electronic sensing can be used for opaque fluids. Knowing the terminal velocity, the size and density of the sphere, and the density of the liquid, Stokes' law can be used to calculate the viscosity of the fluid. A series of steel ball bearings of different diameters are normally used in the classic experiment to improve the accuracy of the calculation. The school experiment uses glycerine or golden syrup as the fluid, and the technique is used industrially to check the viscosity of fluids used in processes. Several school experiments often involve varying the temperature and/or concentration of the substances used in order to demonstrate the effects this has on the viscosity. Industrial methods include many different oils, and polymer liquids such as solutions.

The importance of Stokes' law is illustrated by the fact that it played a critical role in the research leading to at least 3 Nobel Prizes.

Stokes' law is important for understanding the swimming of microorganisms and sperm; also, the sedimentation of small particles and organisms in water, under the force of gravity.

In air, the same theory can be used to explain why small water droplets (or ice crystals) can remain suspended in air (as clouds) until they grow to a critical size and start falling as rain (or snow and hail). Similar use of the equation can be made in the settlement of fine particles in water or other fluids.

Terminal Velocity of Sphere Falling in a Fluid

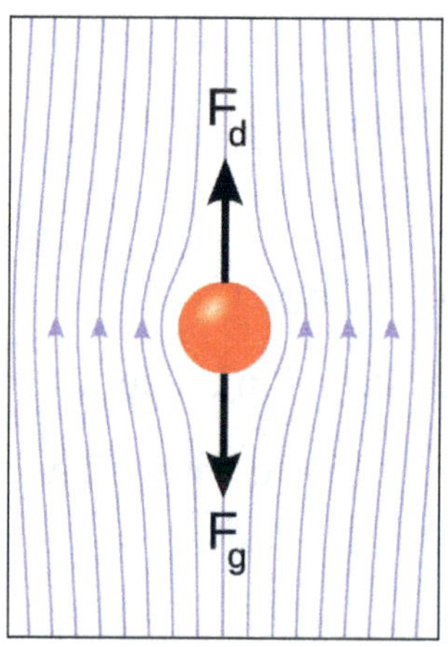

Creeping flow past a falling sphere in a fluid (e.g., a droplet of fog falling through the air): streamlines, drag force F_d and force by gravity F_g.

At terminal (or settling) velocity, the excess force F_g due to the difference between the weight and buoyancy of the sphere (both caused by gravity) is given by,

$$F_g = (\rho_p - \rho_f) g \frac{4}{3} \pi R^3,$$

with ρ_p and ρ_f the mass densities of the sphere and fluid, respectively, and g the gravitational acceleration. Requiring the force balance $F_d = F_g$ and solving for the velocity V gives the terminal velocity V_s. Note that since buoyant force increases as R^3 and Stokes' drag increases as R, the terminal velocity increases as R^2 and thus varies greatly with particle size as shown below. If the particle is falling in the viscous fluid under its own weight, then a terminal velocity, or settling velocity, is reached when this frictional force combined with the buoyant force exactly balances the gravitational force. This velocity V (m/s) is given by:

$$V = \frac{2}{9} \frac{(\rho_p - \rho_f)}{\mu} g R^2$$

(vertically downwards if $\rho_p > \rho_f$, upwards if $\rho_p < \rho_f$), where:

- g is the gravitational acceleration (m/s²).
- ρ_p is the mass density of the particles (kg/m³).
- ρ_f is the mass density of the fluid (kg/m³).
- μ is the dynamic viscosity (kg/m*s).

Steady Stokes Flow

In Stokes flow, at very low Reynolds number, the convective acceleration terms in the Navier–Stokes equations are neglected. Then the flow equations become, for an incompressible steady flow:

$$\nabla \mathrm{p} = \mu \nabla^2 \mathrm{u} = -\mu \nabla \times \omega, \nabla \cdot \mathrm{u} = 0,$$

where:

- p is the fluid pressure (in Pa),
- u is the flow velocity (in m/s), and
- ω is the vorticity (in s⁻¹), defined as $\grave{u} = \nabla \times \mathbf{u}$.

By using some vector calculus identities, these equations can be shown to result in Laplace's equations for the pressure and each of the components of the vorticity vector:

$$\nabla^2 \omega = 0 \text{ and } \nabla^2 p = 0.$$

Additional forces like those by gravity and buoyancy have not been taken into account, but can easily be added since the above equations are linear, so linear superposition of solutions and associated forces can be applied.

Flow Around a Sphere

For the case of a sphere in a uniform far field flow, it is advantageous to use a cylindrical coordinate system (r , φ , z). The z–axis is through the centre of the sphere and aligned with the mean flow direction, while r is the radius as measured perpendicular to the z–axis. The origin is at the sphere centre. Because the flow is axisymmetric around the z–axis, it is independent of the azimuth φ.

In this cylindrical coordinate system, the incompressible flow can be described with a Stokes stream function ψ, depending on r and z:

$$u_z = \frac{1}{r}\frac{\partial \psi}{\partial r}, \qquad u_r = -\frac{1}{r}\frac{\partial \psi}{\partial z},$$

with u_r and u_z the flow velocity components in the r and z direction, respectively. The azimuthal velocity component in the φ–direction is equal to zero, in this axisymmetric case. The volume flux, through a tube bounded by a surface of some constant value ψ, is equal to $2\pi\,\psi$ and is constant.

For this case of an axisymmetric flow, the only non-zero component of the vorticity vector ω is the azimuthal φ–component ω_φ

$$\omega_\varphi = \frac{\partial u_r}{\partial z} - \frac{\partial u_z}{\partial r} = -\frac{\partial}{\partial r}\left(\frac{1}{r}\frac{\partial \psi}{\partial r}\right) - \frac{1}{r}\frac{\partial^2 \psi}{\partial z^2}.$$

The Laplace operator, applied to the vorticity ω_φ, becomes in this cylindrical coordinate system with axisymmetry:

$$\nabla^2 \omega_\varphi = \frac{1}{r}\frac{\partial}{\partial r}\left(r\frac{\partial \omega_\varphi}{\partial r}\right) + \frac{\partial^2 \omega_\varphi}{\partial z^2} - \frac{\omega_\varphi}{r^2} = 0.$$

From the previous two equations, and with the appropriate boundary conditions, for a far-field uniform-flow velocity u in the z–direction and a sphere of radius R, the solution is found to be,

$$\psi = -\frac{1}{2}\mathrm{u}\mathrm{r}^2\left[1 - \frac{3}{2}\frac{\mathrm{R}}{\sqrt{\mathrm{r}^2+\mathrm{z}^2}} + \frac{1}{2}\left(\frac{\mathrm{R}}{\sqrt{\mathrm{r}^2+\mathrm{z}^2}}\right)^3\right]$$

The viscous force per unit area σ, exerted by the flow on the surface on the sphere, is in the z–direction everywhere. More strikingly, it has also the same value everywhere on the sphere:

$$\sigma = \frac{3\mu \mathrm{u}}{2\mathrm{R}}\mathrm{e}_\mathrm{z}$$

with e_z the unit vector in the z–direction. For other shapes than spherical, σ is not constant along the body surface. Integration of the viscous force per unit area σ over the sphere surface gives the frictional force F_d according to Stokes' law.

Faxén's Law

In fluid dynamics, Faxén's laws relate a sphere's velocity **U** and angular velocity **Ù** to the forces, torque, stresslet and flow it experiences under low Reynolds number (creeping flow) conditions.

First Law

Faxen's first law was introduced in 1922 by Swedish physicist Hilding Faxén, who at the time was active at Uppsala University, and is given by,

$$\mathbf{F} = 6\pi\mu a\left[\left(1+\frac{a^2}{6}\nabla^2\right)\mathbf{u}' - (\mathbf{U}-\mathbf{u}^\infty)\right],$$

where

- $\mathbf{F}$ is the force exerted by the fluid on the sphere.
- μ is the Newtonian viscosity of the solvent in which the sphere is placed.
- a is the sphere's radius.
- $\mathbf{U}$ is the (translational) velocity of the sphere.
- $\mathbf{u}'$ is the disturbance velocity caused by the other spheres in suspension (not by the background impressed flow), evaluated at the sphere centre.
- $\mathbf{u}^\infty$ is the background impressed flow, evaluated at the sphere centre (set to zero in some references).

It can also be written in the form,

$$\mathbf{U}-\mathbf{u}^\infty = \mathbf{u}' + b_0\mathbf{F} + \frac{a^2}{6}\nabla^2\mathbf{u}',$$

where $b_0 = -\frac{1}{6\pi\mu a}$ is the mobility.

In the case that the pressure gradient is small compared with the length scale of the sphere's diameter, and when there is no external force, the last two terms of this form may be neglected. In this case the external fluid flow simply advects the sphere.

Second Law

Faxen's second law is given by,

$$\mathrm{T} = 8\pi\mu a^3\left[\frac{1}{2}(\nabla\times\mathbf{u}') - (\Omega-\Omega^\infty)\right],$$

where

- T is the torque exerted by the fluid on the sphere.
- Ù is the angular velocity of the sphere.
- $Ù^{\infty}$ is the angular velocity of the background flow, evaluated at the sphere centre (set to zero in some references).

'Third Law'

Batchelor and Green derived an equation for the stresslet, given by

$$\mathsf{S} = \frac{10}{3}\pi\mu a^3\left[-2\mathsf{E}^{\infty} + \left(1 + \frac{1}{10}a^2\nabla^2\right)\left(\nabla\mathbf{u}' + (\nabla\mathbf{u}')^{\mathrm{T}}\right)\right],$$

where

- S is the stresslet (symmetric part of the first moment of force) exerted by the fluid on the sphere,
- $\nabla\mathbf{u}'$ is the velocity gradient tensor; ^{T} represents transpose; and so $\frac{1}{2}\left[\nabla\mathbf{u}' + (\nabla\mathbf{u}')^{\mathrm{T}}\right]$ is the rate of strain, or deformation, tensor.
- $\mathsf{E}^{\infty} = \frac{1}{2}\left[\nabla\mathbf{u}^{\infty} + (\nabla\mathbf{u}^{\infty})^{\mathrm{T}}\right]$ is the rate of strain of the background flow, evaluated at the sphere centre (set to zero in some references).

Note there is no rate of strain on the sphere (no $\mathbf{E}$) since the spheres are assumed to be rigid.

Faxén's law is a correction to Stokes' law for the friction on spherical objects in a viscous fluid, valid where the object moves close to a wall of the container.

Hagen–Poiseuille Equation

In nonideal fluid dynamics, the Hagen–Poiseuille equation, also known as the Hagen–Poiseuille law, Poiseuille law or Poiseuille equation, is a physical law that gives the pressure drop in an incompressible and Newtonian fluid in laminar flow flowing through a long cylindrical pipe of constant cross section. It can be successfully applied to air flow in lung alveoli, for the flow through a drinking straw or through a hypodermic needle. It was experimentally derived independently by Jean Léonard Marie Poiseuille in 1838 and Gotthilf Heinrich Ludwig Hagen, and published by Poiseuille in 1840 and 1846.

The assumptions of the equation are that the fluid is incompressible and Newtonian; the flow is laminar through a pipe of constant circular cross-section that is substantially longer than its diameter; and there is no acceleration of fluid in the pipe. For velocities and pipe diameters above a threshold, actual fluid flow is not laminar but turbulent, leading to larger pressure drops than calculated by the Hagen–Poiseuille equation.

Equation

Standard Fluid-dynamics Notation

In standard fluid-dynamics notation:

$$\Delta P = \frac{8\mu LQ}{\pi r^4},$$

or

$$\Delta \mathrm{P} = \frac{128\mu \mathrm{LQ}}{\pi \mathrm{d}^4},$$

or

$$\Delta P = \frac{32\mu Lv}{d^2},$$

where:

ΔP is the pressure reduction,

L is the length of pipe,

μ is the dynamic viscosity,

Q is the volumetric flow rate,

r is the pipe radius,

d is the pipe diameter,

π is the mathematical constant π,

v is the velocity.

Physics Notation

$$\Phi = \frac{dV}{dt} = v\pi R^2 = \frac{\pi R^4}{8\eta}\left(\frac{-\Delta P}{\Delta x}\right) = \frac{\pi R^4}{8\eta}\frac{|\Delta P|}{L},$$

where in compatible units (e.g., SI):

Φ is the volumetric flow rate (denoted as Q above),

$V(t)$ is the volume of the liquid transferred as a function of time, t,

v is mean fluid velocity along the length of the tube,

x is distance in direction of flow,

R is the internal radius of the tube,

ΔP is the pressure difference between the two ends,

η is the dynamic fluid viscosity,

L is the length of the tube.

The equation does not hold close to the pipe entrance.

The equation fails in the limit of low viscosity, wide and/or short pipe. Low viscosity or a wide pipe may result in turbulent flow, making it necessary to use more complex models, such as Darcy–Weisbach equation. If the pipe is too short, the Hagen–Poiseuille equation may result in unphysically high flow rates; the flow is bounded by Bernoulli's principle, under less restrictive conditions, by

$$\Phi_{\max} = \pi R^2 \sqrt{2\Delta P / \rho}.$$

Relation to Darcy–Weisbach

Normally, Hagen-Poiseuille flow implies not just the relation for the pressure drop, above, but also the full solution for the laminar flow profile, which is parabolic. However, the result for the pressure drop can be extended to turbulent flow by inferring an effective turbulent viscosity in the case of turbulent flow, even though the flow profile in turbulent flow is strictly speaking not actually parabolic. In both cases, laminar or turbulent, the pressure drop is related to the stress at the wall, which determines the so-called friction factor. The wall stress can be determined phenomenologically by the Darcy–Weisbach equation in the field of hydraulics, given a relationship for the friction factor in terms of the Reynolds number. In the case of laminar flow:

$$\Lambda = \frac{64}{\mathrm{Re}}, \quad \mathrm{Re} = \frac{2\rho v r}{\eta},$$

where Re is the Reynolds number, ρ is the fluid density, v is the mean flow velocity, which is half the maximal flow velocity in the case of laminar flow. It proves more useful to define the Reynolds number in terms of the mean flow velocity because this quantity remains well defined even in the case of turbulent flow, whereas the maximal flow velocity may not be, or in any case, it may be difficult to infer. In this form the law approximates the *Darcy friction factor*, the *energy (head) loss factor*, *friction loss factor* or *Darcy (friction) factor* Λ in the laminar flow at very low velocities in cylindrical tube. The theoretical derivation of a slightly different form of the law was made independently by Wiedman in 1856 and Neumann and E. Hagenbach in 1858 (1859, 1860). Hagenbach was the first who called this law the Poiseuille's law.

The law is also very important in hemorheology and hemodynamics, both fields of physiology.

The Poiseuilles' law was later in 1891 extended to turbulent flow by L. R. Wilberforce, based on Hagenbach's work.

Derivation

The Hagen–Poiseuille equation can be derived from the Navier–Stokes equations. Although more lengthy than directly using the Navier–Stokes equations, an alternative method of deriving the Hagen–Poiseuille equation is as follows.

Liquid Flow Through a Pipe

Assume the liquid exhibits laminar flow. Laminar flow in a round pipe prescribes that there are a bunch of circular layers (lamina) of liquid, each having a velocity determined only by their radial distance from the center of the tube. Also assume the center is moving fastest while the liquid touching the walls of the tube is stationary (due to the no-slip condition).

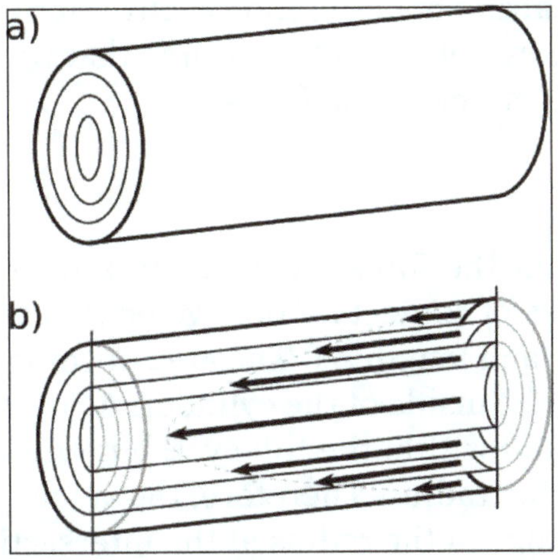

a) A tube showing the imaginary lamina. b) A cross section of the tube shows the lamina moving at different speeds. Those closest to the edge of the tube are moving slowly while those near the center are moving quickly.

To figure out the motion of the liquid, all forces acting on each lamina must be known:

1. The pressure force pushing the liquid through the tube is the change in pressure multiplied by the area: $F = -A\Delta P$. This force is in the direction of the motion of the liquid. The negative sign comes from the conventional way we define $\Delta P = P_{end} - P_{top} < 0.$.
2. Viscosity effects will pull from the faster lamina immediately closer to the center of the tube.
3. Viscosity effects will drag from the slower lamina immediately closer to the walls of the tube.

Viscosity

Two fluids moving past each other in the *x* direction. The liquid on top is moving faster and will be pulled in the negative direction by the bottom liquid while the bottom liquid will be pulled in the positive direction by the top liquid.

When two layers of liquid in contact with each other move at different speeds, there will be a shear force between them. This force is proportional to the area of contact *A*, the velocity gradient in the direction of flow $\Delta v_x / \Delta y$, and a proportionality constant η (viscosity) and is given by,

$$F_{viscosity,\ top} = -\eta A \frac{\Delta v_x}{\Delta y}.$$

The negative sign is in there because we are concerned with the faster moving liquid (top in figure), which is being slowed by the slower liquid (bottom in figure). By Newton's third law of motion, the force on the slower liquid is equal and opposite (no negative sign) to the force on the faster liquid. This equation assumes that the area of contact is so large that we can ignore any effects from the edges and that the fluids behave as Newtonian fluids.

Faster Lamina

Assume that we are figuring out the force on the lamina with radius r. From the equation above, we need to know the area of contact and the velocity gradient. Think of the lamina as a ring of radius r, thickness dr, and length Δx. The area of contact between the lamina and the faster one is simply the area of the inside of the cylinder: $A = 2\pi r\Delta x$. We don't know the exact form for the velocity of the liquid within the tube yet, but we do know (from our assumption above) that it is dependent on the radius. Therefore, the velocity gradient is the change of the velocity with respect to the change in the radius at the intersection of these two laminae. That intersection is at a radius of r. So, considering that this force will be positive with respect to the movement of the liquid (but the derivative of the velocity is negative), the final form of the equation becomes,

$$F_{\text{viscosity, fast}} = -2\pi r\eta\Delta x \left.\frac{dv}{dr}\right|_{r}$$

where the vertical bar and subscript *r* following the derivative indicates that it should be taken at a radius of r.

Slower Lamina

Next let's find the force of drag from the slower lamina. We need to calculate the same values that we did for the force from the faster lamina. In this case, the area of contact is at *r+dr* instead of *r*. Also, we need to remember that this force opposes the direction of movement of the liquid and will therefore be negative (and that the derivative of the velocity is negative).

$$F_{\text{viscosity, slow}} = 2\pi(r+dr)\eta\Delta x \left.\frac{dv}{dr}\right|_{r+dr}$$

Putting it all Together

To find the solution for the flow of a laminar layer through a tube, we need to make one last assumption. There is no acceleration of liquid in the pipe, and by Newton's first law, there is no net force. If there is no net force then we can add all of the forces together to get zero,

$$0 = F_{\text{pressure}} + F_{\text{viscosity, fast}} + F_{\text{viscosity, slow}}$$

or

$$0 = \Delta P 2\pi r dr - 2\pi r\eta\Delta x \left.\frac{dv}{dr}\right|_{r} + 2\pi(r+dr)\eta\Delta x \left.\frac{dv}{dr}\right|_{r+dr}$$

First, to get everything happening at the same point, use the first two terms of a Taylor series expansion of the velocity gradient:

$$\left.\frac{dv}{dr}\right|_{r+dr}=\left.\frac{dv}{dr}\right|_{r}+\left.\frac{d^2v}{dr^2}\right|_{r}dr.$$

The expression is valid for all laminae. Grouping like terms and dropping the vertical bar since all derivatives are assumed to be at radius *r*,

$$0=\Delta \mathrm{P}2\pi \mathrm{r}\mathrm{dr}+2\pi\eta \mathrm{dr}\Delta \mathrm{x}\frac{\mathrm{dv}}{\mathrm{dr}}+2\pi \mathrm{r}\eta \mathrm{dr}\Delta \mathrm{x}\frac{\mathrm{d}^2\mathrm{v}}{\mathrm{dr}^2}+2\pi\eta(\mathrm{dr})^2\Delta \mathrm{x}\frac{\mathrm{d}^2\mathrm{v}}{\mathrm{dr}^2}$$

Finally, put this expression in the form of a differential equation, dropping the term quadratic in *dr*.

$$-\frac{1}{\eta}\frac{\Delta P}{\Delta x}=\frac{d^2v}{dr^2}+\frac{1}{r}\frac{dv}{dr}$$

It can be seen that both sides of the equations are negative: there is a drop of pressure along the tube (left side) and both first and second derivatives of the velocity are negative (velocity has a maximum value at the center of the tube, where $r = 0$). Using the product rule, the equation may be re-arranged to:

$$-\frac{1}{\eta}\frac{\Delta \mathrm{P}}{\Delta \mathrm{x}}=\frac{1}{\mathrm{r}}\frac{\mathrm{d}}{\mathrm{dr}}\left(\mathrm{r}\frac{\mathrm{dv}}{\mathrm{dr}}\right).$$

The right-hand side is the radial term of the Laplace operator ∇^2, so this differential equation is a special case of the Poisson equation. It is subject to the following boundary conditions:

$v(r)=0$ at $r=R$ -- "no-slip" boundary condition at the wall

$\frac{dv}{dr}=0$ at $r=0$ -- axial symmetry.

Axial symmetry means that the velocity v(r) is maximum at the center of the tube, therefore the first derivative $\frac{dv}{dr}$ is zero at $r = 0$.

The differential equation can be integrated to:

$$v(r)=-\frac{1}{4\eta}r^2\frac{\Delta P}{\Delta x}+A\ln(r)+B.$$

To find A and B, we use the boundary conditions.

First, the symmetry boundary condition indicates:

$$\frac{dv}{dr}=-\frac{1}{2\eta}r\frac{\Delta P}{\Delta x}+\frac{A}{r}=0 \text{ at } r=0.$$

A solution possible only if A = 0. Next the no-slip boundary condition is applied to the remaining equation:

$$v(R) = -\frac{1}{4\eta}R^2\frac{\Delta P}{\Delta x} + B = 0$$

so therefore

$$\mathrm{B} = \frac{1}{4\eta}\mathrm{R}^2\frac{\Delta \mathrm{P}}{\Delta \mathrm{x}}.$$

Now we have a formula for the velocity of liquid moving through the tube as a function of the distance from the center of the tube,

$$v = \frac{1}{4\eta}\frac{\Delta P}{\Delta x}(R^2 - r^2)$$

or, at the center of the tube where the liquid is moving fastest (r = 0) with R being the radius of the tube,

$$v_{max} = \frac{1}{4\eta}\frac{\Delta P}{\Delta x}R^2.$$

Poiseuille's Law

To get the total volume that flows through the tube, we need to add up the contributions from each lamina. To calculate the flow through each lamina, we multiply the velocity (from above) and the area of the lamina.

$$\Phi(r)dr = \frac{1}{4\eta}\frac{|\Delta P|}{\Delta x}(R^2 - r^2)2\pi r dr = \frac{\pi}{2\eta}\frac{|\Delta P|}{\Delta x}(rR^2 - r^3)dr$$

Finally, we integrate over all lamina via the radius variable r.

$$\Phi = \frac{\pi}{2\eta}\frac{|\Delta P|}{\Delta x}\int_0^R (rR^2 - r^3)dr = \frac{|\Delta P|\pi R^4}{8\eta\Delta x}$$

Poiseuille's Equation for Compressible Fluids

For a compressible fluid in a tube the volumetric flow rate and the linear velocity is not constant along the tube. The flow is usually expressed at outlet pressure. As fluid is compressed or expands, work is done and the fluid is heated or cooled. This means that the flow rate depends on the heat transfer to and from the fluid. For an ideal gas in the isothermal case, where the temperature of the fluid is permitted to equilibrate with its surroundings, and when the pressure difference between ends of the pipe is small, the volumetric flow rate at the pipe outlet is given by,

$$\Phi = \frac{dV}{dt} = v\pi R^2 = \frac{\pi R^4 (P_i - P_o)}{8\eta L} \times \frac{P_i + P_o}{2P_o} = \frac{\pi R^4}{16\eta L}\left(\frac{P_i^2 - P_o^2}{P_o}\right)$$

where:

P_i inlet pressure.

P_o outlet pressure.

L is the length of tube.

η is the viscosity.

R is the radius.

V is the volume of the fluid at outlet pressure.

v is the velocity of the fluid at outlet pressure.

This is usually a good approximation when the flow velocity is less than mach 0.3

This equation can be seen as Poiseuille's law with an extra correction factor $\frac{P_i + P_o}{2} \times \frac{1}{P_o}$ expressing the average pressure relative to the outlet pressure.

Electrical Circuits Analogy

Electricity was originally understood to be a kind of fluid. This hydraulic analogy is still conceptually useful for understanding circuits. This analogy is also used to study the frequency response of fluid-mechanical networks using circuit tools, in which case the fluid network is termed a hydraulic circuit.

Poiseuille's law corresponds to Ohm's law for electrical circuits, $V = IR$.

Since the net force acting on the fluid is equal to,

$$\Delta F = S\Delta P,$$

where $S = \pi r^2$, i.e. $\Delta F = \pi r^2 \Delta P$, then from Poiseuille's law,

$$\Delta P = \frac{8\mu LQ}{\pi r^4}$$

it follows that

$$\Delta F = \frac{8\mu LQ}{r^2}.$$

For electrical circuits,let n be the concentration of free charged particles,(in m^{-3}) and let q^* be the charge of each particle (in coulombs). (For electrons, $q^* = e = 1{,}6.10^{-19} C$.)

Then nQ is the number of particles in the volume Q, and nQq^* is their total charge. This is the charge that flows through the cross section per unit time, i.e. the current I. Therefore, $I = nQq^*$.

Consequently, $Q = \frac{I}{nq^*}$, and

$$\Delta F = \frac{8\mu LI}{nr^2 q^*}$$

But $\Delta F = Eq,$, where q is the total charge in the volume of the tube. The volume of the tube is equal to $\pi r^2 L$, so the number of charged particles in this volume is equal to $n\pi r^2 L$, and their total charge is,

$$q = n\pi r^2 L q^*.$$

Now,

$$\mathrm{E} = \frac{\Delta \mathrm{F}}{\mathrm{q}} = \frac{8\mu \mathrm{I}}{\mathrm{n}^2 \pi \mathrm{r}^4 (\mathrm{q}^*)^2}.$$

Since the voltage $V = EL,$, we get

$$V = \frac{8\mu LI}{n^2 \pi r^4 (q^*)^2}$$

This is exactly Ohm's law, where the resistance $R = \frac{V}{I}$ is described by the formula,

$$R = \frac{8\mu L}{n^2 \pi r^4 (q^*)^2}..$$

It follows that the resistance R is proportional to the length R of the resistor, which is true. However, it also follows that the resistance R is inversely proportional to the fourth power of the radius r, i.e. the resistance R is inversely proportional to the second power of the cross section area $S = \pi r^2$ of the resistor, which is wrong (according to the electrical analogy)!

The correct relation is,

$$R = \frac{\rho L}{S},$$

where ρ is the specific resistance; i.e. the resistance R is inversely proportional to the first power of the cross section area S of the resistor.

The reason why Poiseuille's law leads to a wrong formula for the resistance R is the difference between the fluid flow and the electric current. Electron gas is inviscid, so its velocity does not depend on the distance to the walls of the conductor. The resistance is due to the interaction between the flowing electrons and the atoms of the conductor. Therefore, Poiseuille's law and the hydraulic analogy are useful only within certain limits when applied to electricity.

Both Ohm's law and Poiseuille's law illustrate transport phenomena.

Medical Applications – Intravenous Access and Fluid Delivery

The Hagen Poiseuille's equation is useful in determining the flow rate of IV fluids that may be achieved using various sizes of peripheral and central cannulas. The equation states that flow rate is proportional to the radius to the fourth power, meaning that a small increase in the internal diameter of the cannula yields a significant increase in flow rate of IV fluids. The radius of IV cannulas is typically measured in "gauge", which is inversely proportional to the radius. Peripheral IV cannulas are typically available (from large to small) 14g, 16g, 18g, 20g, 22g, 24g. As an example, the flow of a 14g cannula is typically twice that of a 16g, and ten times that of a 20g. It also states that flow is inversely proportional to length, meaning that longer lines have lower flow rates. This is important to remember as in an emergency, many clinicians would intuitively favour a long central venous catheter over a short peripherally sited cannula. While of less clinical importance, the change in pressure can be used to speed up flow rate by pressurizing the bag of fluid, squeezing the bag, or hanging the bag higher from the level of the cannula. It is also useful to understand that viscous fluids will flow slower (e.g. blood transfusion).

References

- Chanson, H. (2009). Applied Hydrodynamics: An Introduction to Ideal and Real Fluid Flows. CRC Press, Taylor & Francis Group, Leiden, The Netherlands, 478 pages. ISBN 978-0-415-49271-3
- Daryn Lehoux (2012). What Did the Romans Know? An Inquiry into Science and Worldmaking. University of Chicago Press (ISBN 9780226471143)
- Oertel, Herbert; Prandtl, Ludwig; Böhle, M.; Mayes, Katherine (2004). Prandtl's Essentials of Fluid Mechanics. Springer. pp. 70–71. ISBN 0-387-40437-6
- Chanson, Hubert (2004). Hydraulics of Open Channel Flow: An Introduction. Butterworth-Heinemann. ISBN 0-7506-5978-5
- Goldstein, Herbert; Poole, Charles P., Jr.; Safko, John L. (2002), Classical Mechanics (3rd ed.), San Francisco, CA: Addison Wesley, pp. 347–349, ISBN 0-201-65702-3
- Toro, E.F. (1999). "Chapter 2. Notions on Hyperbolic PDEs". Riemann Solvers and Numerical Methods for Fluid Dynamics. Springer-Verlag. ISBN 3-540-65966-8
- Tipler, Paul (1991). Physics for Scientists and Engineers: Mechanics (3rd extended ed.). W. H. Freeman. ISBN 0-87901-432-6
- Morin, David (2008). Introduction to Classical Mechanics. New York: Cambridge University Press. ISBN 9780521876223
- Feynman, R.P.; Leighton, R.B.; Sands, M. (1963). The Feynman Lectures on Physics. ISBN 0-201-02116-1., Vol. 2, §40–3, pp. 40–6 – 40–9
- Lamb, H. (1994). Hydrodynamics (6th ed.). Cambridge University Press. ISBN 978-0-521-45868-9. Originally published in 1879, the 6th extended edition appeared first in 1932

Permissions

All chapters in this book are published with permission under the Creative Commons Attribution Share Alike License or equivalent. Every chapter published in this book has been scrutinized by our experts. Their significance has been extensively debated. The topics covered herein carry significant information for a comprehensive understanding. They may even be implemented as practical applications or may be referred to as a beginning point for further studies.

We would like to thank the editorial team for lending their expertise to make the book truly unique. They have played a crucial role in the development of this book. Without their invaluable contributions this book wouldn't have been possible. They have made vital efforts to compile up to date information on the varied aspects of this subject to make this book a valuable addition to the collection of many professionals and students.

This book was conceptualized with the vision of imparting up-to-date and integrated information in this field. To ensure the same, a matchless editorial board was set up. Every individual on the board went through rigorous rounds of assessment to prove their worth. After which they invested a large part of their time researching and compiling the most relevant data for our readers.

The editorial board has been involved in producing this book since its inception. They have spent rigorous hours researching and exploring the diverse topics which have resulted in the successful publishing of this book. They have passed on their knowledge of decades through this book. To expedite this challenging task, the publisher supported the team at every step. A small team of assistant editors was also appointed to further simplify the editing procedure and attain best results for the readers.

Apart from the editorial board, the designing team has also invested a significant amount of their time in understanding the subject and creating the most relevant covers. They scrutinized every image to scout for the most suitable representation of the subject and create an appropriate cover for the book.

The publishing team has been an ardent support to the editorial, designing and production team. Their endless efforts to recruit the best for this project, has resulted in the accomplishment of this book. They are a veteran in the field of academics and their pool of knowledge is as vast as their experience in printing. Their expertise and guidance has proved useful at every step. Their uncompromising quality standards have made this book an exceptional effort. Their encouragement from time to time has been an inspiration for everyone.

The publisher and the editorial board hope that this book will prove to be a valuable piece of knowledge for students, practitioners and scholars across the globe.

Index

www.ingramcontent.com/pod-product-compliance
Lightning Source LLC
LaVergne TN
LVHW080705210926
848485LV00003B/421